THE ROLE OF LEGUMES IN THE FARMING SYSTEMS OF THE MEDITERRANEAN AREAS

DEVELOPMENTS IN PLANT AND SOIL SCIENCES

VOLUME 38

The Role of Legumes in the Farming Systems of the Mediterranean Areas

Proceedings of a Workshop on the Role of Legumes in the Farming Systems of the Mediterranean Areas UNDP/ICARDA, Tunis, June 20–24, 1988

Edited by

A.E. OSMAN, M.H. IBRAHIM and M.A. JONES

The International Center for Agricultural Research in the Dry Areas (ICARDA), Aleppo, Syria

KLUWER ACADEMIC PUBLISHERS

DORDRECHT / BOSTON / LONDON

Library of Congress Cataloging-in-Publication Data

The Role of legumes in the farming systems of the Mediterranean areas : proceedings of a workshop on the role of legumes in the farming systems of the Mediterranean areas, UNDP/ICARDA, Tunis, June 20-24, 1988 / edited by A.E. Osman, M.M. Ibrahim, and M.A. Jones.
p. cm. -- (Developments in plant and soil sciences ; 38)
ISBN-13:978-94-010-6949-6 e-ISBN-13:978-94-009-1019-5
DOI: 10.1007/978-94-009-1019-5

1. Legumes--Mediterranean Region--Congresses. 2. Legumes as feed--Mediterranean Region--Congresses. 3. Legumes as food--Mediterranean Region--Congresses. 4. Agricultural systems--Mediterranean Region--Congresses. I. Osman, A. E. II. Ibrahim, M. M. III. Jones, M. A. IV. United Nations Development Programme. V. International Center for Agricultural Research in the Dry Areas. VI. Series: Developments in plant and soil sciences ; v. 38.
SB203.3.M43R65 1990
338.1'733'091822--dc20 89-15633

ISBN-13:978-94-010-6949-6

Published by Kluwer Academic Publishers,
P.O. Box 17, 3300 AA Dordrecht, The Netherlands.

Kluwer Academic Publishers incorporates
the publishing programmes of Martinus Nijhoff,
Dr W. Junk, D. Reidel, and MTP Press.

Sold and distributed in the U.S.A. and Canada
by Kluwer Academic Publishers,
101 Philip Drive, Norwell, MA 02061, U.S.A.

In all other countries, sold and distributed
by Kluwer Academic Publishers Group,
P.O. Box 322, 3300 AH Dordrecht, The Netherlands

printed on acid free paper

Softcover reprint of the hardcover 1st edition 1990

Table of Contents

Foreword

Legumes are an important source of protein for humans and animals. They provide nutritionally rich crop residues for animal feed, and play a key role in maintaining the productivity of soils particularly through biological nitrogen fixation. They are, therefore, of immense value in rainfed farming systems.

The International Center for Agricultural Research in the Dry Areas (ICARDA) has a responsibility for research on food, pasture, and forage legumes. The Center also has the broad objective of improving livestock production in rainfed farming systems. Although food legumes have been known and grown by farmers in the WANA region for a long time, their productivity has remained low and variable. Forage legumes, on the other hand, are not so well known by farmers of the region, and their role in the farming systems is not so well understood. Thus, we need to develop the concept of using forage legumes as crops and to fit them into cropping systems. In its efforts to increase the productivity of food legumes and develop the legume-based crop/livestock systems, ICARDA has established a network of scientists in the different National Agricultural Research Systems in the region. To further strengthen this network, ICARDA convened a workshop on 'The Role of Legumes in the Farming Systems of Mediterranean Areas' in Tunis, Tunisia, 20–24 June 1988. This workshop was co-sponsored by UNDP, who also contributed funds for this publication.

An overview of the role of legumes in the farming systems of the Mediterranean basin itself, and of other parts of the world with a Mediterranean climate, formed the background on which specific topics and discussions were based. The papers presented in the workshop, the discussions, and recommendations are presented in these proceedings which I trust will contribute to a better understanding of the past, present and future role of legumes in farming systems. I am confident that the knowledge, interest and ideas generated by this workshop will assist in developing sound future strategies which will improve the welfare of the less privileged farmers in the Mediterranean areas of the WANA region and elsewhere.

Nasrat R. Fadda
Director General
ICARDA

Acknowledgements

We wish to convey our sincere appreciation to all those who contributed to these proceedings: the authors, the workshop participants, Ahmed Kamel and his staff at the ICARDA Tunis Office, and the secretaries and artists at ICARDA headquarters, especially Aida Battikha, Mary Bogharian, Amira Diab and Abdul-Rahman Hawa. We are especially grateful to Parvin Damania for her invaluable secretarial expertise at a critical time.

The workshop could not have been held nor the proceedings printed without the financial assistance of UNDP and the hospitality of the Government of Tunisia. We thank them for their generosity.

The Editors

PART I

A GLOBAL AND REGIONAL OVERVIEW OF THE ROLE OF LEGUMES IN FARMING SYSTEMS

Legumes in Farming Systems in Mediterranean Climates

I.W. BUDDENHAGEN

Agronomy Department, University of California, Davis, CA 95616, USA

Abstract. Agriculture in the Mediterranean region contrasts sharply from agriculture in the other four distant regions with Med climates, due to its ancient nature in ancient cultures. All winter food and forage legumes of major utility evolved in the Mediterranean/West Asian region but have become more important outside the region. The Mediterranean legumes can become much more useful in their homelands with greater research and development effort focused especially on nitrogen budgets in agrosystems combined with watershed management to reduce overall ecosystem degradation. The major opportunity is in developing an integration with cereals and animals, in a rainfed key system patterned after the Australian experience. Intercropping of food legumes with cereals spaced at lower density in appropriate strips should be explored. Research on the nitrogen budget in all systems needs more focus on actual gains to either inter or rotational crops that really are attributable to N-fixation by the legume. Small gains can be made under irrigation for the higher priced grain legumes and for alfalfa and berseem with further breeding for disease/pest resistance, N-fixation rates and yield. Biological value of food and feed grain legume protein is low and there are many toxicities also. Improvement is needed through better supported and better integrated breeding programs involving different disciplines. This is especially important if any winter grain legume is to compete in stability of yield and in quality in the large markets for animal rations.

Prologue

The Mediterranean region has rightly been called the first 'Eden'. Early Man found it a paradise in climate and interlacings of land and water. With his low population in antiquity there was abundant game and seafood and an easy primitive agriculture (Attenborough 1987).

The Mediterranean region and areas just to the east were cradles of evolving civilization and of the domestication of major animal and many crop species. Our modern civilized world and the radical changes of the last hundred years are unthinkable without the heritage of the religions and

A.E. Osman et al. (eds.), The Role of Legumes in the Farming Systems of the Mediterranean Areas, 3–29.

civilizations that were developed in the Mediterranean region and then passed northward into Europe proper.

The great antiquity of Man and agriculture in the Mediterranean region contrasts markedly with the complete absence, until recently, of civilization and agriculture in the four other widely separated regions of the Earth having 'Mediterranean' climates. In fact, the other four areas of Mediterranean (hereinafter called Med) climates were either largely uninhabited or were utilized sparsely by fisher and hunter folk until around 1750–1850 AD. They were then colonized by Spaniards, English, or Dutch, who brought livestock and crops from their homelands. The crops brought from the Mediterranean flourished and the annual grass weeds accompanying the livestock quickly became established in California and Australia and radically changed the rangeland ecosystem.

Introduction

This overview deals with both forage and cool season food legumes as they are utilized in agricultural systems in the five widely separated regions of the world having Med climates. The food legumes covered are the cool season ones: chickpeas, lentils, fabas and peas. But it should also be recognized that, under irrigation, summer food-legumes are also grown in such climates; these being common beans, lima beans, cowpea, and, to a lesser extent, mung beans, groundnuts, and soybeans.

Briefly covered is lupin, a cool season food legume of considerable promise as a feed grain and well adapted to the Med climates.

Although we are interested in legumes, it should be recognized at the outset that the winter cereals dominate in such climates and that many other crops can be grown, especially with irrigation, and that the food legumes are likely to remain a small crop in relation to total production and area. Moreover, in economic competition, it must be appreciated that the cool season legumes can be, and are, grown in areas more temperate than the Med climates. Thus, in Europe, North and South America, Australia, and Asia, chickpeas, peas, fabas, and lentils are grown outside the Med climates in areas of colder winters, shifting there to spring-summer crops because of the more severe winter temperatures.

Legume forages are many, but in Med climates the annual clovers and medics (*Medicago* spp.), along with alfalfa under irrigation, dominate. The prime center for research on legume forages in recent years has been in Australia, working with species derived from the Mediterranean region. North America has been the major center for research and development of alfalfa. In recent years legume forages have grown in importance and acreage, especially in Australia and California, and there is renewed interest in them in the Mediterranean region.

The Mediterranean Climates

The Med climates can be defined as having annual rainfall of 275–900 mm, with more than 65% falling in winter. At least one winter month must have an average temperature below 15°C but yearly hours below O°C should not exceed 3% of the total (Aschmann 1973a). In layman's terms, Med climates have dry summers and mild wet winters (most of the rain falls in autumn and spring). The summers range from cool along the western coasts to hot inland and in more easterly regions (Fig. 1).

Within this broadly defined frame, the Med climates have considerable diversity which affects dryland crop and forage potential. For example,

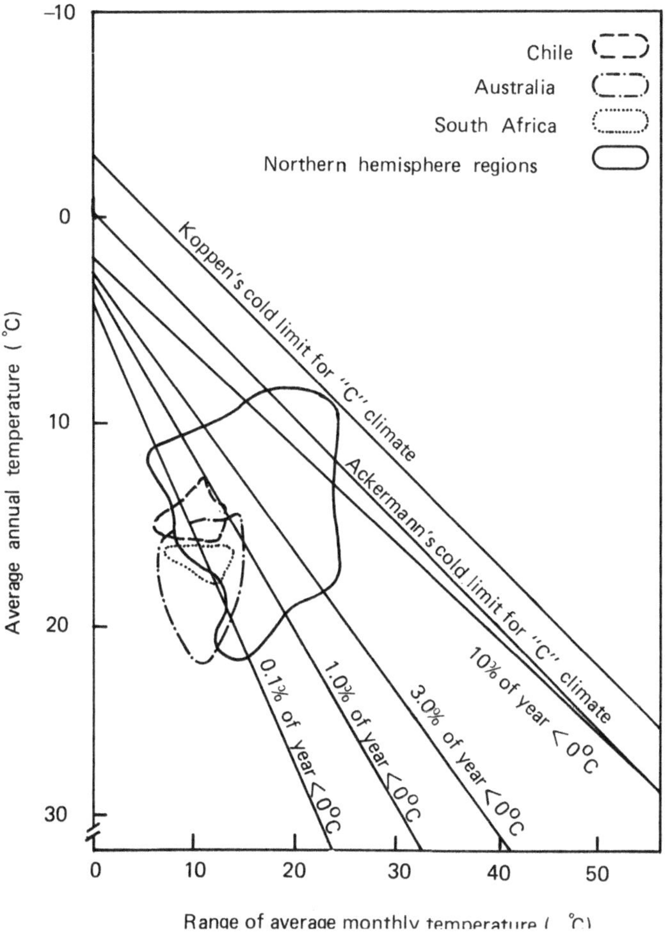

Fig. 1. Thermal characteristics of Mediterranean regions as defined by Koppen.

summer rainfall may range from zero to 35% of a total which may itself range from 275 to 900 mm. Both winter and summer temperatures may vary widely within the defined limits, and evapotranspiration (ET), although usually very high, may be low along fogbound western coasts. The rugged lands in most Med climates create altitude gradients and local microclimates which vary greatly from place to place, affecting crop species and variety adaptation and potential.

But everywhere, the dominating aspect of Med climates on agriculture is the paucity and erratic nature of rainfall and the coincidence of drought and heat during summer. In the drier areas this results in an inversion of the normal growing season, with the rest period in summer rather than, or in addition to, the winter.

Sunshine is uniformly high, above 3000 hours per year, with little year to year variation. This creates high potential evapotranspiration (PET), which greatly exceeds rainfall during most of the year. PET is so great that under plant cover, soil water reserves in spring are soon exhausted.

The dominating aspect on both crop and natural plant growth becomes the changing yearly pattern of soil water availability as PET and temperature interact with rainfall, absence of rainfall, plant cover and soil water-holding and delivery capacity. In Med climates, water deficit is the key limitation to productivity, and its amelioration through cropping pattern adjustments or by irrigation and water conservation manipulation is the key to increased productivity (Arnon 1979).

Winds, sometimes violent, also occur in many Med climates, adding to the variability of climate and its impact on agriculture.

The Med Climate Lands

For those of us living in Med climates it is hard to realize how unique and how limited such regions are on earth. Also, most of us working within one Med ecosystem seldom consider those remote other Med areas – so distantly dispersed on the earth's surface.

Less than 2% of the earth's habitable land surface has Med climates and these are scattered in five widely separated regions between 32°–4° latitude: The Mediterranean, Southwestern Africa, Southern Australia, Chile and California (Thrower and Bradbury 1973). Thus, there are two Northern and three Southern Hemisphere Med climate areas (actually four, since Med climates occur in two separated areas in Australia: South West and South Central).

The Med climates lie on the western or southwestern edges of continents and around the Mediterranean Sea. Three regions (California, Chile, SW Africa) have offshore cold currents which contribute to the formation of Med climates eastward.

The Mediterranean region has over half the total world Med climate land area and it differs in several ways from the others, not only in its diversity but in its ancient human and agricultural history.

Pertinent to consideration of agriculture in Med climates are the regions bordering them which offer potential for the same crops. Where they grade into climates too dry for the Med climate label, as in N. Africa, and parts of Australia they may be transformed into highly productive lands with Med crops by irrigation. Med climates also grade into other climates by becoming too wet or by having more summer rain, as in Southern Europe and elsewhere and yet remain appropriate for the Med-adapted food and forage legumes. Med climates may grade into steppe because they are too cold in winter in more continental or mountainous areas, such as in Iran and Washington State (USA), and yet these remain ideal for the Med-adapted legumes, if not too cold, or if their cultivation is shifted from winter-spring to spring-summer.

Ecology and Evolution

There are many treatments of the ecology and evolution of Med climate plants and animals (di Castri and Mooney 1973). Good comparative reviews of plant evolution and natural plant productivity exist for native ecosystems in the widely separated Med climates (Raven 1973; Axelrod 1973). In perusing this literature one is struck by how carefully the ecologists avoid a consideration of agriculture, and even of the effects of Man and his animals on the ecosystem. There are some exceptions, which are well worth reading (Aschmann 1973b) where a few ecologists have emphasized environmental degradation by Man and his animals in the Mediterranean region itself. In actual fact, Man and his animals have had and continue to have a major effect on soil, plant biota, evolution, and productivity in these most fragile ecosystems.

From the ecological/evolutionary literature, however, there are some key points for crop and agroecosystem researchers:

- Mediterranean climates are only very recent (appearing in the Pleistocene) and floras in these climates are of recent evolution, from subhumid to semiarid plant communities at the margin of the tropics, and also in the Northern Hemisphere, from temperate Tertiary floras (Raven 1973).
- In the Southern Hemisphere the very different floras in each Med area evolved in isolation and mainly from tropical border lands, tracing to an original Mesozoic southern continent.
- In each Med area about 10% of the genera and at least 40% of species are endemic.
- The Med climate floras, which evolved independently from different ancestors, converged in adaptive traits in remarkably similar ways:

1) All areas support a similar-appearing 'sclerophyll' shrubby flora and much work has been carried out to characterize the sclerophyll habit's adaptation to Med climate stresses.
2) The major recent evolution has been the expansion of annual plant species, both grasses and forbs, which in the Northern Hemisphere Med climates, comprise about 50% of total species. This annual habit is of most interest to us since this ultimate adaptation to summer-drought, through seed-escape and a cool-season-growth-habit has given Man his most important cereal and several very important food and forage legume species.
3) A third adaptation is summer survival of perennials through large roots or crowns or other subterranean organs, such as in many Liliaceae and in other families. Alfalfa probably fits here in evolutionary adaptation, even though its origin is further east, in a steppe climate.
4) A fourth adaptation of large perennials to summer drought is to lose leaves through a summer-shedding habit or for leaves to die and remain hanging.

Origin and Domestication of the Legumes Now in Med Climates

The cool-season food legumes belong to three tribes, the Vicieae (*Lens*, *Pisum*, *Vicia faba*, *Lathyrus*), the recently separated tribe Cicereae (*Cicer*), and the Genisteae (*Lupinus*) (Summerfield and Bunting 1980; Duke 1981; Hebblethwaite 1983). All are of Mediterranean and West Asian derivation and domestication, with Turkey to Iran as major centers (Simmonds 1976).

Several genera extend to East Africa (*Lens*, *Lupinus*, *Lathyrus*). *Lupinus* has evolved separately also in the Americas, where the major center of its diversity exists, and where it was separately domesticated ('Tarwi' or Peru) (Gladstones 1980; Pate et al. 1985). Species of *Vicia* and *Lathyrus* also occur indigenously in the Americas.

All of the cool-season food legumes are ancient domesticates in the lands from Iran-Iraq to the Mediterranean and most were associated, along with the primitive wheats and barley, with the rise of civilizations in the Eastern Mediterranean and Fertile Crescent. Peas (Davies et al. 1985) may be from that area but they are now mostly grown in more northerly climates. Many found their way to India, which became a secondary center of diversity, especially for *Cicer* (Smithson et al. 1985). Lupins remained most restricted, probably due to their greater toxicity; but other species also contain antimetabolitic toxins, especially *Lathyrus* and *V. faba* (Bond et al. 1985). Thus, no cool-season food legume was domesticated or utilized in other Med climate lands, and even the genera were largely missing, except for wild *Lupinus* in California.

The important forage or pasture legumes of Med climate lands and neighboring steppe climates have originated in the Eastern Mediterranean and neighboring Western Asian areas. This region is important for the genera *Trifolium*, *Medicago*, *Vicia*, *Lupinus*, *Lotus*, and three lesser known: *Hedysarum*, *Onobrychis*, and *Ornithopus* (Polhill and van der Maesen 1985). The sweet clovers (*Melilotus spp.*) come from further north. Tropical America is also a primary center of many genera of forage legumes, but these are largely adapted and used in the tropics or subtropics in areas with summer rainfall.

Although species of *Lupinus*, *Trifolium*, and *Lotus* are indigenous to the California Med climate, none has proven very productive when confronted with modern grazing practices. In Chile, the African Cape, and Australia, indigenous forage legumes were either missing, or, like in California, they have not evolved with much grazing pressure and they cannot compete with the Mediterranean introductions.

There are, however, 'browse' legumes, of more diverse origin, native in all Med climate lands, which have value both as browse plants and in nitrogen fixation as part of the nitrogen cycle in nature. These shrubs and small trees have considerable utility in poor agrarian Med climate areas and the recent interest in such plants is a welcome addition to our traditional crop orientation.

The oldest 'domesticated' fodder legume is alfalfa (*Medicago sativa*), used since ancient times, along with berseem (*Trifolium alexandrinum*) and white clover (*T. pratense*) (Hanson et al. 1988; Mannetje et al., 1980). There seems to be no great antiquity in the tradition of sown pasture legumes.

Legumes were included in crop rotations in Europe to 'improve' the soil, with no understanding of why until the mid nineteenth century. Clovers then became much more important in permanent pastures and interest and research continues to expand, driven by the importance of N-fixation and the need for N replacement in cropped agroecosystems.

There is a need for low cost N on millions of hectares of poor rangelands as well as on older cropped soils where N has been depleted. Likewise, there is a need for high protein fodder everywhere, which can be met by including legumes in a grass pasture or fodder mix. Thus, there is great interest in forage legumes as a means of economically maintaining nitrogen balance in both primitive and advanced agroecosystems.

Adaption and Productivity

Since Med climates are summer-dry climates and also rainfall-variable climates, plants have evolved diverse adaptations for survival of the stresses imposed by the high PET conditions of hot and dry summers. Four plant strategies were briefly mentioned in a previous section.

A more critical analysis, however, reveals that the requirement of agriculture is not adaptation through evolution for survival but rather, productivity. Productivity in Med climates requires the efficient use of the limited water available (improved water use efficiency) (Elston and Bunting 1980). A second means to increase productivity is to increase, through cropping and soil management practices, the amount of water that can be utilized by the crop.

The aim should be to obtain as much growth as possible during the cool season when PET is low and rainfall is probable. It is here where an increase in evapotranspiration (ET) should result in increased growth without adversely affecting the overall water balance. Thus, we should capitalize on the evolutionary strategy, whether of annuals or perennials, wherein growth occurs in both autumn and spring, during cold, cool, and warm days.

For winter annuals already evolved for high seed yield as biomass is quickly converted to abundant seed set before summer death, we need to improve the crop's ability to develop and accumulate biomass during cool weather and, by increasing sink size as well as by increasing soil water extraction ability as summer approaches, to match grain fill with terminal water extraction (Buddenhagen and Richards 1988). This may require not just increasing winter hardiness, but adding to cool-weather adaptation for growth. How much genetic variation there is within species for this latter characteristic is hardly researched. Additionally, to raise productivity of food legumes, higher levels of disease and insect resistance are needed since cool-rainy-season diseases and insects often occur which were avoided earlier by growing the crop in summer when rain is slight.

The traditional conversion of chickpeas to a spring-sown crop from an autumn germinating ancestor appears to have occurred in the Mediterranean region to escape epidemics of *Ascochyta* blight, a rain-splash favored disease (Saxena and Singh 1987). In California, where *Ascochyta* blight is absent, the crop became a spring-sown crop apparently as a means to escape serious aphid-borne virus diseases, largely vectored from February through April (Bosque-Perez et al. 1988).

For annual forages we may wish to increase total biomass accumulation and even *reduce* seed yield if this occurs at the expense of usable protein biomass and if the seed is lost to the grazing system, and is in excess of regeneration needs. We also may wish to increase the period of spring-early summer growth through deeper root systems, or altered photoperiod phenology.

For perennial legumes we need consider only some pasture and forage legumes, especially various clovers, medics, vetches, and *Hedysarum*. For all of the legume crops, increasing cool season productivity probably also means improving *Rhizobium*-host relations for N-fixation at cooler temperatures.

Soil Variability, Constraints, and Degradation

Soils develop through time from parental rocks under the influences of climate, vegetation, soil animal and microbe biota, relief and aspect.

Although very similar Med climates occur in widely separated regions, the soils differ markedly, largely due to differences in parental rock, time of exposure, and relief. They also differ markedly *within* each Med climate area.

In general, the Mediterranean area is underlain by sedimentary rocks that are calcareous: limestones, marls, and siliceous sandstones. Med lands elsewhere in the world are underlain more by igneous and metamorphic rocks and their soils in general are less basic. But everywhere there is considerable soil variability, both in class, depth, clay type and content, waterholding capacity, cation exchange capacity, and fertility (Roquero 1979).

To the crop-focused biological scientist, it is difficult to appreciate the overriding effects of soil properties in determining crop productivity and crop sustainability (Munns and Mosse 1980). Since soil varies from farm to farm and valley to valley to hillside slope, crop output also varies and crop breeding objectives and agronomic recommendations shift from appropriate to inappropriate. Often, the relevance of experiment station trials to farm reality is minimal because of differences in soil type, soil depth, and recent differences in soil/crop/input histories between the farmers' fields and the research station.

Residual soils predominate in Med climates and these are usually deficient in nitrogen and phosphorous, especially where there is a long history of cropping, as in the Mediterranean. These may also be shallow, eroded, or stony. Hilly lands, grazed and often burned, as in the California foothills, are often deficient in sulfur as well (Jones 1979). In Australia, minor element deficiencies occur on old residual soils. Acidification occurs in residual Med soils in Australia and California with more intensive cropping and nitrogen, P, and S inputs (Munns 1986). Most of the Mediterranean has ample calcium carbonate or other bases and acidification is not a problem.

In all Med regions many residual soils lack the depth, clay content, and structure which enable ideal permeability and water holding capacity for high crop productivity under the existing rainfall regimes. Organic matter in such soils is low and its contribution is mainly in improving soil physical properties. Improved cropping systems which enhance organic matter levels offer some opportunities for fertility improvement through tighter nutrient recycling.

To the practical agriculturalist in Med climates the major concern regarding soils is their capacity to infilter and store the limited rainfall, and to deliver it to crop root systems (Damagnez 1979). As these capacities become limited, erosion and degradation increase, further limiting nutrient reserves, which are already deficient in nitrogen, and if long cropped, in phosphorus and possibly other elements.

In contrast to residual soils are the alluvial soils of Med climates, which are often of considerable depth and water-holding capacity. In the great California valleys, such alluvial soils, derived from recently exposed rocks in nearby mountains, are both deep and nutrient rich. They offer enormous productive potential given the sunny Med climate, especially with irrigation, but intensive cropping for only a few decades reveals the absolute need for nitrogen,and lesser needs of P, K, and Zn. However, irrigation in Med climates can soon lead to salinity and alkalinity problems.

In such deep soils, winter legume crops could have considerable potential for low input farming since they will largely fix their own nitrogen (Larson et al. 1988). However, these crops are grown on only a small area due to low world prices in relation to yield levels and high overhead costs. For winter food legumes, only chickpea is considered economic in California at present, where it needs one or two irrigations. On rich alluvial soils where irrigation systems exist, the legumes are generally displaced by crops with higher profit-margin potential. The major exception on a large scale in California is alfalfa, grown for seed, hay, and ensilage on 0.5 M hectares, as a high yielding summer irrigated and spring-summer-growing perennial. Ninety percent is grown for hay (7.5 t/ha) and ten percent is either cubed or made into silage; pasturing is rarely done. Some 30,000 ha are cut for seed, averaging 300 kg/ha. Here also, summer food legumes are grown under irrigation on a small scale, a scale which saturates the market, as presently envisaged, for cowpeas (*Vigna unguiculata*), common beans (*Phaseolus vulgaris*), and lima beans (*P. lunatus*). Berseem (*Trifolium alexandrinum*), an annual summer forage, already grown extensively on valuable land in Egypt under irrigation, is being newly revived as a potential crop for California (Graves et al. 1987).

The Nitrogen Budget and Related Issues

The great interest in legumes is stimulated firstly by their capacity to fix nitrogen. Thus, in chronically nitrogen-deficient soils, this attribute appears, on the surface, to be a 'godsend', especially in rotations with cereals. But the situation is not so simple, as will be elaborated below.

The second stimulus for growing legumes is their high protein levels, either in forage for livestock or in seeds for human or animal food. In overpopulated regions, and for poor people generally, where little meat is eaten, having food-legumes available at hand or at low cost is, indeed, very important. In richer or meat-oriented societies, food legumes are less important and could be dispensed with entirely. In such societies and areas, forage legumes assume more importance, since they can help produce lower cost meat. A more rare attribute of interest, and only for summer legumes, is that of oil food/feed value, as in soybeans and groundnuts.

The forage legumes offer an additional attribute – that of potentially enhancing rangeland productivity indirectly by enhancing grass growth (through N-fixation and transfer), thus providing greater ground cover and lower erosion.

In many natural ecosystems (sans Man) legumes must be very important in relation to the nitrogen cycle and budget, but there are few data to support this assumption. Also, we should acknowledge the hundreds of millions of years of plant and animal life on earth before legumes evolved.

Our concern, however, is legumes in farming systems in Med climates; the nitrogen fixation part of this subject was reviewed in an ICARDA-sponsored workshop in early 1986 (Beck and Materon, 1988). The many papers in that volume and the literature in general reveal the complexity of N-fixation and the many gaps between laboratory data and field reality. Many basic questions still have no answers, or the answers are equivocal. Much research information is available but few field activities exist utilizing the information or proving or disproving its validity on a large scale in the real economic world (Anon. 1975; Tainton 1981).

The major exception is in Australia, where research on forage legumes has led to their use on a large scale (12 million hectares) in a managed system of ley farming with improved varieties, *Rhizobium*-inoculated seed, and P inputs (Webber et al. 1977; White et al. 1978; Cocks et al. 1980; Puckridge and French 1983).

One additional key to understanding establishment and maintenance of legumes in mixtures is their need for light, revealed in two classical papers by Stern and Donald (1962a,b). To appreciate this need for light in relation to relative height of a shading competitor, such as grass for subclover, or a cereal and an interplanted short clover or medic, would be important to designing useful cropping system experiments.

The other exception to research payoff is in California and elsewhere, where alfalfa growing is carried out with improved cultivars and management practices (Hanson et al. 1988).

We speak of the legumes as of great importance because of their N-fixing capacity. But in the Mediterranean area, grain legumes are grown on only 3.5% of cropland, compared with 53% for cereals. Although forage legumes exist there over vast areas, they are generally at low populations in natural, unimproved rangelands. Their contribution to the N-budget is probably much less than many think. Where hay is prepared for winter in the cold Med-fringe climates (central Turkey, etc.) the vetches are often raised as an intercrop with oats or barley and there a real N-input is probably added to the system by fixation (Acikgoz 1988).

But in general, in relation to the whole agroecosystem in the Mediterranean region, and in California, I believe the role of legumes in N-fixation is generally overstated and overthought. One reason is that the researcher, with his clear figures of N-fixation, sees an obvious advantage to legumes and predicts great benefits. Unfortunately, he does not have to make a living in crop farming or animal farming, and thus, the practical decision making the farmer must do to survive is not part of the researchers' equations. As a Cypriot researcher wrote recently, 'What will happen if production of faba bean increases? Most grain legumes are consumed by humans and generally the possibility of increasing local consumption is rather limited' (Papastylianou 1988).

All of this is not to deny the present reality or the potential of legumes in agriculture. The four major cool season food legumes are grown on ±2.6 million ha in the Mediterranean/Middle East and on ±23 million ha worldwide (Table 1) (FAO 1985). Legume forages, either as sole or mixed crops, are grown in a managed way on probably more than 100 M ha worldwide. Alfalfa alone is on 33 M ha. Subclovers and medics in Australia are planted on 12 M ha. The countries and areas of greatest production of both the forages and winter food legumes are all outside the Med climates (Table 2).

Some simple and important questions can be asked about N-fixation of legumes in Med climates under existing farm practices:

Table 1. Trends in cool season food legume statistics.*

Regions/Crops	Area (1000 ha) 1979–81	Area (1000 ha) 1985	Yield (kg/ha) 1979–81	Yield (kg/ha) 1985	Prod. (1000 MT) 1979–81	Prod. (1000 MT) 1985
Faba bean						
USA/Chile/S.Af.	–	–	–	–	–	–
Australia	11	17	531	639	6	10
Med/ME[a](16 c.)	721	714	1098	1173	790	897
World	3680	3155	1164	1321	4283	4170
Lentil						
Australia	–	–	–	–	–	–
U.S.A.	78	41	1076	911	84	37
Chile	50	36	505	679	25	25
South Africa	–	–	–	–	–	–
Med/ME (14 c.)	493	931	779	783	425	763
World	1902	2469	601	668	1142	1650
Dry Pea						
Australia	46	147	1175	1143	55	168
U.S.A.	70	89	2310	1751	163	155
Chile	17	10	747	850	13	9
South Africa	8	4	938	1250	8	5
Med/ME (12 c.)	125	134	1306	1408	104	114
World	7470	8948	1172	1301	8718	1164
Chickpea						
Australia	–	–	–	–	–	–
Mexico	195	150	1119	1000	218	150
Chile	18	11	511	813	9	9
South Africa/USA	–	–	–	–	–	–
Med/ME (18 c.)	641	836	848	851	560	707
World	9951	9512	627	675	5985	6416

[a] Med/ME = Mediterranean – Middle East Region.

* 1985 FAO Production Yearbook, Vol. 39.

Table 2. Countries with largest production (1000 t) of cool season food legumes.*

Crop	Largest Producer In world	In Med.
Lentil	India (544)	Spain (51)
		Syria (48)
Faba	China (2300)	Egypt (307)
Pea	China (1900)	Iran (50)
Chickpea	India (4547)	Turkey (380)

* FAO Production Yearbook, 1985, Vol. 39.

Cool Season Food Legumes raised for Grain

Does the amount of seed removal plus stover result in a negative, zero, or positive nitrogen balance on the land utilized for production?

The answer to this question is not clear. Many studies have been conducted to estimate the total N fixed by food legumes in different places and under different conditions (Roughley 1980). In general, the data indicate that 50–80% of the plant biomass N comes from fixation. There are inherent problems with methodology in measuring N fixation (Witty and Minchin 1988). But even more frustrating is that seldom are calculations made in relation to actual N removed from the field in the crop products.

Thus, under actual existing agricultural practices it is difficult to know the N budget for grain legumes. If seed-N equals fixed-N then the N in the rest of the plant represents N from soil cycling, etc., and removal of the plants in hand-harvesting methods or removing them for livestock feeding will result in a negative N balance for the field. (Grazing will conserve 90% of the N eaten, through urine and dung deposition, but this will be erratic spatially). An ensuing cereal crop should have less N than if the land had been fallowed. (There may or may not be other benefits of a grain-legume crop to the following cereal crop, as measured by yield changes over a control, that are not N related). The basic point here is that if there is no extra residual N, one cannot use the argument of N-fixation benefits beyond the legume crop itself; other reasons for advocating its use in a cropping sequence must be advanced.

Results from rotation yield trials usually indicate a slightly higher yield of a cereal following a food legume as compared with cereal after cereal, or sometimes, even after a fallow (Saxena 1988). It is reasonable to assume that N is largely responsible for the yield jump compared with cereal-after cereal since the legume crop uses less soil N. Yield comparisons with after-fallow are not so clear since the fallow should conserve the soil N. Without N-budget work it is not clear that fixation *per se* is the cause of higher cereal yield after legume than after fallow. Many other reasons could be advanced. Other than disease levels, one could hypothesize that the legume is sequestering nitrate that would otherwise be leached. In a recent study in California with winter lupin it was calculated that seed N exceeded the quantity of N fixed by 23–68 kg/ha (Larson et al. 1988).

It seems that both short and long term rotational yield trials in different environments with different food legumes and different legume product extractions are needed. These must have an N-budget component, using the best methodology. Then for each region a reasonable calculation should be made of the proportion of cereal production that could have a food legume rotation, based on actual markets and economic considerations.

If one wishes to advocate intercropping cereals and food legumes for the reasons of benefit from N-fixation, there would have to be N leakage of N that would otherwise be lost. It is not at all clear that this would be the case. Thus, one needs proof; lacking it prevents certainty that any extra cereal production is attributable to the N-fixation reason.

In any case it seems reasonable that more research should be conducted on intercropping cereals and food legumes, in various potentially useful configurations (especially in narrow strips) to determine if potential advantages might occur in total yield, whether due to N-fixation or disease reduction or whatever. Researchers should be willing to back away from dense sowings in such experiments, and to emphasize and evaluate the economic and practical considerations.

Forage Legumes in Farming Systems

Four situations should be considered:

a) Annually sown and harvested annual species (vetches, peas, berseem clover);
b) Perennials largely cut for hay or silage with little grazing [alfalfa, sulla (*Hedysarum coronarium*)];
c) Pasture legumes which are longer-term in the farming systems in rotation with cereals (Cocks 1988), which are usually reseeding annual, but sometimes perennial, species (alfalfa, sulla, annual medics, subclover); and
d) Seeding legumes into natural rangelands where, combined with inputs, they are intended to increase animal productivity in a profitable way.

The first system is where the annually sown legumes are traditionally used for hay or as harvested green forage, but they may be grazed profitably (Cocks 1988). Their influence on a subsequent cereal grain yield is usually positive compared with cereal after cereal, but not always compared with cereal-after fallow. Here again it would be useful to know what portion of cereal yield increase, if any, is due to earlier N-fixation. It would also be of interest to project on a larger scale, local potential amount and consequences of a shift from fallow-cereal to forage legume-cereal. Major increases in meat production in the Mediterranean and West Asia would require export opportunities in order not to lower prices drastically.

This cereal-annually harvested forage legume rotation was formerly very common in the USA and elsewhere outside of Med climates, in colder winter areas. In California it is now little-practiced and harvested forage is mainly from pure stands of perennial, irrigated alfalfa (the second system). In this system alfalfa lasts about four years before a rotational crop is grown and it fixes a great deal of N. But the N is mostly exported with the hay, as well as substantial amounts of K (680 kg/ha in four years of harvest).

In Australia, ley farming (the third system) has been developed on a large scale, and this system is now being tried in Morocco (Ismaili and Bentassil 1988) and elsewhere in the Mediterranean. This system alternates winter-spring annual legume pasture with cereal production in cycles of short but differing lengths. In Australia this system supposedly results in a net N gain of 50 kg/ha/yr under subclover (Russell 1960). The essence of the system is to interpose a winter wheat crop into a legume or legume/grass pasture which recycles from self sown seed in the wet cool season which follows the wheat year.

Although the Australian system has been a boon to agriculture in Med climates there, and it is now being advocated in the Mediterranean region, it appears not to be a stable system (Stern 1985; Davidson 1987). A major problem in Australia appears to be in the acidification of the soils by the N-fixing system. The legume content of their pastures is less than that of 25 years ago, and is disturbingly low (Stern 1985). Some factors are the increasing acidification, a decline in soil structure with increased runoff, and a decline in phosphate fertilizer use due to its increasing costs. (Phosphate additions are essential for maintaining the legume in the system). Many questions remain inadequately answered, including the real contribution of the legume to the overall short- and long-term N budget.

In spite of these present problems in Australia, it would appear that many of the soils in the Mediterranean area would not soon suffer acidification. Moreover, it would appear that good local research in the Mediterranean region targeted directly at bringing about change in farming practices by legume-cereal rotations would have a great beneficial payoff. Probably the best approach is to follow the Australian lead in concepts, but use local germplasm selected from relevant local ecosystems, and make other changes based on the local situation.

The fourth system, that of adding legumes to natural rangelands with the idea of permanent improvement without rotations, has been developed in California. The basic approach is to clear the rolling hills in patchwork fashion of the sclerophyllous brush, by fire or mechanically, and to sow inoculated subclover or medics with P and S (Graves et al. 1987; Jones et al. 1981, 1983). Much research has shown the benefits of this approach but even in California the adoption rate has been low. There appear to be subtle economic and operational limitations which also involve the extensive nature of livestock operations (Menke 1988). To adopt this approach in the Mediterranean and West Asia would require a very holistic approach and full appreciation of local traditions.

Pathogens, Pests, Symbionts, Co-evolution, 'Adaptation', and Breeding

A fundamental point rarely emphasized is the co-evolutionary background of our crops and animals with their pathogens, pests, and symbionts. For the legumes evolving in the Med climates it is clear that the co-evolutionary relationship in the Mediterranean and West Asia, combined with the antiquity of domestication and agriculture in that region, dominate the recent development of these crops in the distant Med climate regions, as well as in more temperate steppe climates in the Americas and elsewhere. As with most of our crops, the great advances in technology, breeding, productivity and farming, have occurred outside their centers of origin, brought about by people of different cultures in lands deficient in plant genetic resources.

The reasons for crop success away from home are diverse but four points stand out:

1) The absence of ancient cultural traditions which accept the status quo, these traditions being common in the local poor rural societies where crop domestication occurred.
2) The original absence in the new lands of many of the damaging pests and pathogens co-evolved with the crop, at first enabling better growth and less pest/pathogen damage in an escape situation. This has been and continues to be followed by re-encounters, with pests and pathogens being introduced from their centers of origin to their old hosts in the new centers of development, often revealing super-susceptibility and stimulating much research, including resistance breeding (Buddenhagen 1977, 1983).
3) For the legumes, the lack of a co-evolved N-fixing symbiont, *Rhizobium*, was a special case. The local rhizobia in new Med climate lands were generally ineffective on the introduced legumes. Some co-evolved rhizobia must have been introduced into the distant Med-climate regions inadvertently with the largely accidental introduction of the seed and straw of the Med legumes themselves. But generally, the soils in the distant lands did not enable effective nodulation, and once the *Rhizobium* strain concept was known it was clear that research and development had to be carried out on strain matching and seed inoculation, to make the new legumes successful in their new homes. (Quality control of *Rhizobium* strains for subclover in California remains a major problem in stand establishment). This stimulus is largely not needed in the co-evolved centers and it is still not clear that research in this area will be very rewarding on food legume production in the centers of origin.

4) The original extensive nature of agriculture in the new lands, with an absence of labor, created a stimulus for mechanization, and the European immigrants soon developed mechanical and other innovations in agriculture, probably also stimulated by an unfettering from the old rigid class societies left behind.

In any event, for the Med legumes in their original homeland, the key to their recent evolution must be grazing pressure combined with co-evolutionary development of effective rhizobia in each soil microclimate. The Mediterranean region offers both acid and basic soils, cold sites and hot and droughty ones, even salinity and alkalinity. Thus, there has been long adaptation, under grazing pressure, for vigor and survival under many different pressures. This grazing pressure adaptation has enabled them to out-compete all indigenous legumes evolved in other Med climates. This special evolution in diverse habitats emphasizes the value of the Mediterranean/Middle East region for genetic resources. Also, it explains the difficulty of increasing yields through seed inoculation, at least in soils where legumes are already abundant and ancient. Long adaptation has also produced the seed burying character of subclover, and the seed dispersal and survival (hardseededness) mechanisms of both medics and subclovers.

So both the needs and potentials of the species differ markedly between ‘homeland’ and ‘new land’. The researcher should take cognizance of this fact and analyze his or her area with an evolutionary eye.

A chronology of significant events for alfalfa in the USA (Barnes et al. 1988) makes the co-evolutionary point and reveals the stimulus of re-encounters on research: 1925 – Bacterial wilt described; 1942 – Release of the first two wilt resistant varieties; 1954 – Discovery in New Mexico of the spotted alfalfa aphid; 1957 – Release of two varieties resistant to spotted alfalfa aphid; late 1960’s – New era of multiple disease and insect resistance begins; late 1960’s – ‘Breakdown’ of resistance to spotted alfalfa aphid by ‘new’ biotypes; 1974 – Blue alfalfa aphid identified in California; 1976 – Verticillium wilt first identified in western USA; 1976 – Release of a blue aphid-resistant alfalfa; and finally, to emphasize how the industry can drive breeding programs into new directions to accomplish specific new goals: 1986 – Release of ‘Nitro’ alfalfa, nondormant, with increased N fixation and storage of N in roots for use in short-term rotations.

Food legumes have been called the step-child of the green revolution because their breeding has not resulted in any major yield gain. Because of this they have been shunted off the best soils as well, in many places. In retrospect, one should not have expected a yield breakthrough as in the small cereals since that breakthrough was really an N-response, one related to short plant stature and harvest index. The food legumes were already there in the sense that they fix their own nitrogen and many had already been selected for determinate growth habit and short stature. In fact, some, such as lentils (Muehlbauer et al. 1985), were inherently too short to start with.

Thus, gains have been small and have been mainly through stabilizing existing yield potential levels by adding disease and insect resistance. Additionally, some small gains have been made in breeding for increased N-fixation (Teuber et al. 1984; Mytton et al. 1988), an attribute which may be more useful in a properly managed cropping system than increasing yields *per se*. All of these gains are related to host/pathogen/pest/symbiont evolutionary and co-evolutionary interactions, and this area requires more basic work for greater progress that is long-lasting (Buddenhagen 1977; Gould 1983).

In general, the instability of yield is a major factor influencing the unpopularity of food legumes in higher-technology agricultural regions. This is especially true for fabas and chickpeas. Disease and insect vulnerability appear to be major problems, and often the diseases are little understood, especially for those caused by viruses and root pathogens. Continuous cropping seems not to work well and thus an integration of these crops in well studied rotations that are stable would appear to be a good research target under local situations, with teamwork by entomologists, pathologists, plant breeders, and agronomists.

The legumes seem low yielding (for seed) in general when compared with the cereals, and the reasons are still not entirely clear. N-fixation costs are considered to consume 10–30% of the photosynthate (Schubert and Ryle 1980), but protein is supposed to take only 0.75 g of sugar carbon per gram of protein synthesized (Norton et al. 1985), so the high protein in seeds should not be inimical to high yield potential. Obviously, more basic research is required on yield potential and metabolic costs and biochemistry of the legumes. Having non-nodulating mutant isolines would help in unraveling this complex area.

Progress in breeding soybeans for increased yield of the high oil-containing seeds (that are produced at high metabolic cost) shows what can be done with greater research effort. Also, for the forage legumes, alfalfa is a model for progress, although its outcrossing breeding system, enabling a recurrent selection, population improvement breeding approach, makes it more akin to maize than to most inbreeding legumes.

Nutritional, Anti-Nutritional Aspects

Legumes are often touted as excellent high-protein foods. Promotional campaigns exist now in the USA, hoping to increase demand for beans by pushing the idea of their high nutritional value. Food legumes are said to be the poor man's meat, implying that they are as good as meat but also implying that they are for the poor. In reality the real implication is that as affluence spreads, demand for food legumes will decrease; indeed, this has already occurred.

But the high protein of food legumes must be looked at in relation to actual biological value of the consumed product. When this is done it is revealed that, collectively, there are various negative nutritional factors in most food legumes, both for humans and for animals (Bressani and Elias 1980). The protein is of poor digestibility with a low biological value and it is deficient in the sulfur-containing amino acids cystine and methionine. Moreover, many food legumes contain anti-nutritional factors of various types (Norton et al. 1985; Bresani and Elias 1980). Even alfalfa has anti-nutritional factors (Howarth 1988).

Those of us who work for greater legume production tend to overvalue legumes as food, probably because lab analyses show high protein levels and we associate high protein with high quality. Actual feeding studies to show biological value are few and are largely published in journals plant breeders and agronomists do not read. Moreover, such studies are site- or culture-specific and to have relevance they must simulate the diets of the consuming public in the area of our interest. Except for soya protein, which has a high value, food legumes are negative in human infant diets and their protein cannot substitute for milk. Surveys have shown that mothers do not feed common beans to small children, much less to infants, because they are poorly tolerated (Norton et al. 1985).

The prime value of food legumes appears to be for poor societies having cereals as staples, since small-to-moderate additions of food legumes in such diets balance the amino acid requirements. In rice diets, maximum *protein* quality values occurred when 80% of the protein came from rice and only 20% from beans or mung (Bressani and Elias 1980). This means a 10:1 ratio of rice: beans. Many studies with maize/bean diets show that the protein ratios should be 1:1 which would mean a raw product ratio of about 4:1 maize: beans. For both diets, however, protein quality and weight gains are increased by amino acid supplements. Possibly such *in vitro* studies have been carried out for chickpeas, fabas, lentils, and peas in relation to cereal based diets in the Mediterranean and Middle East, but most of the literature appears to be on common bean. Where diets are primarily root-based starches, addition of food legumes does not provide a balancing of amino acid needs, and protein digestibility remains low.

Low protein digestibility is caused by various factors: tannins, trypsin inhibitors, haemagglutinin compounds, protein tertiary structure, and cell walls. It has been calculated that for beans (*P. vulgaris*) with a 64% protein digestibility and a 1000 kg/ha yield, only 147 kg of protein would be utilized by human consumption of that one hectare total yield. This represents a waste of all the protein in 360 kg of that 1000 kg bean yield (Bressani and Elias 1980).

In addition to all this, there are specific anti-nutritional compounds in certain species: cyanogens, anti-vitamin E factors, alkaloids, flavones and isoflavones, toxic amino acids, saponins and pyrimidine glucosides (Nowacki 1980). Some of these occur in forages and affect their nutritional value as well. For instance, the toxic saponin, medicagenic acid, is present in alfalfa and medics; and flavones and isoflavones which are estrogenic and reduce lambing rates are found in subclover and other legumes, especially when infected by fungi. The diseases favism and lathyrism result from eating seeds of those crops, which contain different toxic pyrimidine glucosides. Legumes in general have evolved many compounds, possibly for pest defense reasons, which affect mammalian digestion and metabolism in various negative ways. They vary from species to species and amongst cultivars.

A positive note is that several food legumes appear to reduce serum cholesterol levels (Norton et al. 1985). Also, they may be useful in regulating diabetes. Research in these areas is in its infancy.

Improving digestibility and reducing toxicities should be prime breeding objectives and this area is both complex and neglected. As breeders attempt to increase pest and disease resistance, they may be increasing product toxicity and reducing its digestibility. Few programs link these two subjects and the need for *in vivo* data rather than *in vitro* data make this area of research difficult and expensive.

It would seem, however, that each important food, feed, and forage legume species should have at least one center, where breeding for improvement includes not only yield and adaptability and disease/pest resistance, but also *nutritional biological quality*, in an integrated program. This is especially important if we wish to project grain legume expansion for product inclusion in feed rations where cooking is not practical to remove the heat-labile (Liener 1980) negative factors.

Mechanization and Equipment, Including Processing

In high-tech agriculture in California and Australia, practical improvements and adaptation of mechanical devices for sowing, applying fertilizers and pesticides, and harvesting are a constant part of evolving agricultural development. The push comes from farmers, agribusiness companies, and machinery making companies. Research backup is carried out by colleges and universities which have departments of mechanical or agricultural engineering. There is a spill-over from needs of other industries into food and forage legumes farming and processing. The subject is driven by needs to cut costs and stay competitive and by opportunities provided by high-quality product output. Any crop soon declines without cost-cutting and quality-improving machines and innovations. The food legumes must use machinery giving spacings of the dominant crop in the rotation (in California, often cotton) and they must use the equipment, or simple modifications of equipment, developed and used for other crops (usually cereals for harvesters). Either direct combining is carried out, or the crop is swathed, wherein it dries, and is combined later with a pick-up head on a self-propelled combine. For crops such as the food legumes, which are basically low yielding, mechanization is essential to keep down unit costs and thus have a profitable product at a reasonable price.

In third world countries this area is generally poorly developed, and product price remains relatively very high in relation to local buying power. This is mainly due to low yields combined with high hand-harvesting costs. The possible solutions would seem to be: 1) raise and stabilize yields (discussed earlier), and 2) trim labor costs to production; especially by developing and using practical cutting and threshing machines.

This latter needs elaboration. Since most winter food legumes are low growing and low podding, mechanical harvesting results in both field loss and picking up of dirt and stones. Several approaches to this problem can be taken. The first is for the breeders to work on a taller, stiffer plant type with podding higher up the stem and less tendency to shatter. The second is to work out locally, land preparation procedures, especially at sowing, to smooth the land surface and to roll any small stones into the soil surface. Cultipackers have been developed which work for this purpose. The third is to develop simple reapers with pickup guards that will 'windrow' the crop. The fourth is to develop a low cost combine (and work out collective ownership) that will pick up and thresh the windrow, or, alternatively cut and thresh in one direct operation. If this latter is not realistic, then a pick-up device can be developed and wagons used to carry the cut material to a threshing site where a low-cost thresher (also to be developed, refined, or promoted) can thresh out the grain.

One point to keep in mind in relation to harvesting and threshing is the utility of the plant biomass in the farming system: is it to be part of the nitrogen and organic matter budget for the next cereal crop and therefore needs to be evenly distributed over the land surface? Or is it to be fed to livestock in a few feed lots. Or can browsing it as an aftermath be carried out over the whole field? These are important considerations to development of lower cost harvesting methods.

For export markets and quality considerations, processing equipment in terms of cleaning and sizing and packaging needs to be considered, to increase local margins on distant final market product.

For forage legumes made into hay, silage, or cubes, the alfalfa innovations in the USA and in California especially, should be consulted (Pauli et al. 1988).

Agroecosystem and Watershed Management Focus and the Future

Farming Systems

Crops are grown in cropping systems and sometimes in farming systems. A farming system has been defined as the *pattern of resources and processes of resource use in a farming unit* (Norman 1979). This treatment by Norman of cropping systems in the tropics is well worth reading by those concerned with systems in the Mediterranean region, since there are many third world resemblances and these areas face similar problems in relation to hydrologic, energetic, biogeochemical, and socioeconomic aspects of agricultural development.

But in most third world countries, dealing with farming systems in relation to the *farming unit* is too restrictive. The farm unit has neighboring units and usually a village orientation. In rough terrain the real unit may be a valley and the surrounding hills, with a village center. Changes can be brought about by gaining data which make holistic sense in relation to a specific agroecosystem, or valley or watershed.

In high tech regions, research and development is highly centered on the crop itself. The cropping system is largely left to the interplay of the educated farmers and their perception of markets and interplay with agribusiness and local, constantly revised experience. Thus, third world students studying in high-tech countries receive very little training appropriate for holistic research targeted at bringing about positive changes at the agroecosystem level in countries with traditional low-tech, low-yielding agriculture.

The bias has been on crop improvement and crop agronomy to the neglect of the cropping system, farming system, and agroecosystem. Indeed, we are educated as discipline people: geneticists, plant pathologists, etc., not as agriculturalists. There is much talk about farming systems research and although trends are in that direction, change comes slowly and organizational problems in research management are considerable in order to bring about useful team-oriented holistic project accomplishment.

Most of the International Centers have wrestled with this problem for several years and excellent work, approaches, and protocols have been published by IRRI, CIAT, ICRISAT, IITA, CIMMYT and ICARDA, which should be very useful for those concerned with farming systems in the Mediterranean and West Asian region. Much is applicable even though most of it is from the tropics. A good case study from Syria has been published (Gibbon and Martin 1978), and there must be many others.

The problem, of course, is that the real need is *both* for good analytical discipline research and good appropriate crop research, and for these to be combined with good analytical holistic farming systems research.

Agroecosystem and Watershed Focus

Beyond a farming systems approach is an orientation to agroecosystem or watershed management. This is even more holistic, and if sites are representative of a large class and are well chosen for development activities and investment, the payoff has great potential.

The reason for this is that group activities can be fostered and a whole village or valley group can be activated to try new innovations, and most importantly, to suggest their own innovations, and in effect, educate the research team in life's economic and practical realities.

Attempts to change the present reality have the best chance of success if they follow a study of historical changes and the evolution of the system one is attempting to influence. There is an excellent historical treatment on Mediterranean Agriculture in Grigg's book (1974) on 'The Agricultural Systems of the World: An Evolutionary Approach'.

The main goal is to develop a sustainable agriculture at a higher level of productivity and profitability for the benefit of the common good. The major constrictions are water shortage and nutrient deficiencies. By focusing holistically on a watershed, with existing knowledge of water harvesting, erosion control, world germplasm, and nutrient cycling, much progress can be made. Legumes can fit well into such research and development projects, but the focus should be on maximizing water conservation and use, and nutrient recycling and conservation for the total system. How to best utilize the special characteristics of legumes, economically, for such larger goals is both our challenge and our opportunity.

A final caveat for us all is the old Spanish proverb 'Del dicho al hecho hay un gran trecho'. – From word to deed there is a long path!

Acknowledgements

The author is grateful for informative discussions on this subject with Drs John Menke, Milt Jones, Walt Graves, Don Munns, Charlie Raguse and Fred Muehlbauer. Also for the loan of slides from Drs Walter Kaiser and Larry Teuber. Special thanks are extended to Dr Nilsa Bosque-Perez for critical discussions and help during manuscript preparation.

References

Acikgoz, E. 1988. Annual forage legumes in the arid and semi-arid regions of Turkey. Pages 47–54 *in* Nitrogen Fixation by Legumes in Mediterranean Agriculture (Beck, D.P. and Materon, L.A. eds) Martinus Nijhoff Publishers, The Netherlands.

Anonymous. 1975. Improvement and utilization of grazing resources in Mediterranean-type climates. Proceedings of an international seminar held at The Regional Centre for Agrarian Research and Development, 'La Orden', Guadajira, Spain.

Arnon, I. 1979. Optimizing yields and water use in Mediterranean agriculture. Pages 311–347 *in* Soils in Mediterranean Type Climates and Their Yield Potential. Proceedings 14th Colloqium International Potash Institute Dev Bund, A.G. Bern, Switzerland.

Aschmann, H. 1973a. Distribution and peculiarity of Mediterranean ecosystems. Pages 11–19 *in* Mediterranean Type Ecosystems: Origin and and Structure (di Castri, F. and Mooney, H.A. eds), Springer-Verlag, New York, USA.

Aschmann, H. 1973b. Man's impact on the several regions with Mediterranean climates. Pages 363–371 *in* Mediterranean Type Ecosystems: Origin and and Structure (di Castri, F. and Mooney, H.A. eds), Springer-Verlag, New York, USA.

Attenborough, D. 1987. The First Eden: The Mediterranean world and man. Little, Brown, and Co. Boston, Mass.

Axelrod, D.I. 1973. History of the Mediterranean ecosystem in California. Pages 225–227 *in* Mediterranean Type Ecosystems: Origin and and Structure (di Castri, F. and Mooney, H.A. eds), Springer-Verlag, New York, USA.

Barnes, D.K., Goplen, B.P. and Baylor, J.E. 1988. Highlights in the U.S.A. and Canada. Pages 1–24 *in* Alfalfa and Alfalfa Improvement (Hanson, A.A., Barnes, D.K. and Hill, R.R. eds). Agronomy Vol.29. ASA, CSSA, SSA, Madison, Wisconsin, USA.

Beck, D.P. and Materon, L.A. (eds). 1988. Nitrogen fixation by legumes in Mediterranean agriculture. Martinus Nijhoff Publishers, The Netherlands.

Bond, D.A., Lawes, D.A., Hawtin, G.C., Saxena, M.C. and Stephens, J.H. 1985. Faba bean. Pages 199–265 *in* Grain legume crops (Summerfield, R.J. and Roberts, E.H. eds). Collins, London, UK.

Bosque-Perez, N.A., Buddenhagen, I.W. and Duffus, J.E. 1988. Virus diseases of chickpea in California. Phytopathology. (In press).

Bressani, R. and Elias, L.G. 1980. Nutritional value of legume crops for humans and animals. Pages 135–155 *in* Forage Legumes for Energy-Efficient Animal Production (Barnes, R.F., Ball, P.R., Brougham, R.W., Marten, G.C. and Minson, D.J., eds) USDA-ARS, USA.

Buddenhagen, I.W. 1977. Resistance and vulnerability of tropical crops in relation to their evolution and breeding. Annals New York Academy of Sciences 287: 309–326.

Buddenhagen, I.W. 1983. Breeding strategies for stress and disease resistance in developing countries. Annual Rev. of Phytopathology 21: 385–409.

Buddenhagen, I.W. and Richards, R.A. 1988. Breeding cool season food legumes for improved performance in stress environments. Pages 81–95 *in* World crops: Cool season food legumes (Summerfield, R.J., ed.). Kluwer Academic Publishers, Dordrecht, The Netherlands.

Cocks, P.S. 1988. The role of pasture and forage legumes in livestock farming systems. Pages 3–10 *in* Nitrogen fixation by legumes in Mediterranean agriculture (Beck, D.P. and Materon, L.A. eds) Martinus Nijhoff Publishers, The Netherlands.

Cocks, P.S., Mathison, M.J. and Crawford, E.J. 1980. From wild plants to pasture cultivars: Annual medics and subterranean clover in Southern Australia. Pages 569–596 *in* Forage Legumes for Energy-Efficient Animal Production (Barnes, R.F., Ball, P.R., Brougham, R.W., Marten, G.C. and Minson, D.J., eds) USDA-ARS, USA.

Damagnez, J. 1979. Mediterranean climates: the economic use of water and crop production. Pages 43–55 *in* Soils in Mediterranean Type Climates and Their Yield Potential. Proceedings 14th Colloquium International Potash Institute Dev Bund, A.G. Bern, Switzerland.

Davidson, S. 1987. Combating soil acidity: three approaches. Rural Research 134: 1–10.

Davies, D.R., Berry, G.J., Heath, M.C. and Dawkins, T.C. 1985. Pea. Pages 147–198 *in* Grain legume crops (Summerfield, R.J. and Roberts, E.H. eds). Collins, London, UK.

di Castri, F. and Mooney, H.A. (eds). 1973. Mediterranean type ecosystems: origin and structure. Springer-Verlag, New York, USA.

Duke, J.A., 1981. Handbook of legumes of world economic importance. Plenum Press, New York, USA.

Elston, J. and Bunting, A.H. 1980. Water relations of legume crops. Pages 37–42 *in* Forage Legumes for Energy-Efficient Animal Production (Barnes, R.F., Ball, P.R., Brougham, R.W., Marten, G.C. and Minson, D.J., eds) USDA-ARS, USA.

FAO Production Yearbook. 1985. Vol.39. FAO, Rome.

Gibbon, D. and Martin, A. 1978. Food legumes in the farming system: A case study from Northern Syria. Pages 223–28 *in* Food Legume Improvement and Development (Hawtin, G.C. and Chancellor, G.I., eds). ICARDA and IDRC.

Gladstones, J.S. 1980. Recent developments in the understanding, improvement and use of *Lupinus*. Pages 603–611 *in* Forage Legumes for Energy-Efficient Animal Production (Barnes, R.F., Ball, P.R., Brougham, R.W., Marten, G.C. and Minson, D.J., eds) USDA-ARS, USA.

Gould, F. 1983. Genetics of plant-herbivore systems: Interactions between applied and basic study. Pages 599–653 *in* Variable Plants and Herbivores in Natural and Managed Systems (Denno, R. and McClure, R., eds). Academic Press, New York, USA.

Graves, W.L., Burgess, L.K., Weitkamp, W.H. and George, M.R. 1987. Hardseeded Spanish subclover finds a place in southern California. California Agriculture 41: 8–9.

Graves, W.L., Williams, W.A., Wegrzyn, V.A., Calderon, D., George, M.R., and Sulling, J.L. 1987. Berseem clover is getting a second chance. California Agriculture 41: 15–18.

Grigg, D.B. 1974. The agricultural systems of the world: An evolutionary approach. Cambridge University Press, Cambridge, UK.

Hanson, A.A., Barnes, D.K. and Hill, R.R. (eds). 1988. Alfalfa and Alfalfa Improvement. Agronomy Vol. 29. ASA, CSSA, SSSA, Madison, Wisconsin, U.S.A.

Hebblethwaite, P.D. 1983. The Faba Bean (*Vicia faba* L.). Butterworths, London, UK.

Howarth, R.E. 1988. Antiquality factors and non-nutritive chemical components. Pages 493–514 *in* Alfalfa and Alfalfa Improvement (Hanson, A.A., Barnes, D.K. and Hill, R.R. eds). Agronomy Vol.29. ASA, CSSA, SSSA, Madison, Wisconsin, U.S.A.

Ismaili, M. and Bentassil, A. 1988. The use of biological nitrogen fixation to develop livestock and cereal production in Morocco. Pages 41–46 *in* Nitrogen Fixation by Legumes in Mediterranean Agriculture (Beck, D.P. and Materon, L.A. eds) Martinus Nijhoff Publishers, The Netherlands.

Jones, M.B. 1979. Biogeochemical cycling in annual grassland ecosystems. The Botanical Review 45: 111–144.

Jones, M.B., Koenigs, R.L., Vaughn, C.E. and Murphy, A.H. 1981. Effect of converting chaparral to grassland on soil fertility in a Mediterranean -type climate. Page 121 *in* XIV International Grassland Congress Summaries (Smith, J. Allan and Hays, V.W. eds) Westview Press, Boulder, Colorado, USA.

Jones, M.B., Williams, W.W. and Vaughn, C.E. 1983. Soil characteristics related to production on subclover-grass range. Journal Range Management 36(4): 444–446.

Larson, K.J., Cassman, K.G. and Phillips, D.A. 1988. Yield, dinitrogen fixation and aboveground N balance of irrigated white lupin in a Mediterranean climate. Submitted for publication (Agronomy Journal).

Liener, I.E. 1980. Heat-labile anti-nutritional factors. Pages 157–169 *in* Forage Legumes for Energy-Efficient Animal Production (Barnes, R.F., Ball, P.R., Brougham, R.W., Marten, G.C. and Minson, D.J., eds) USDA-ARS, USA.

Mannetje, L.'t, O'Connor, K.F. and Burt, R.L. 1980. The use and adaptation of pasture and fodder legumes. Pages 537–551 *in* Forage Legumes for Energy-Efficient Animal Production (Barnes, R.F., Ball, P.R., Brougham, R.W., Marten, G.C. and Minson, D.J., eds) USDA-ARS, USA.

Menke, J.W. 1988. Management controls on productivity. *In* The California Annual Grassland Ecosystem (Mooney, H.A. and Huenneke, L.F., eds). In press.

Muehlbauer, F.J., Cubero, J.I. and Summerfield, R.J. 1985. Lentil. Pages 266–311 *in* Grain legume crops (Summerfield, R.J. and Roberts, E.H. eds). Collins, London, UK.

Munns, D.N. 1986. Acid soil tolerance in legumes and *Rhizobia*. Pages 63–91 *in* Advances in Plant Nutrition. Vol.2. (Tinhs, B. and Laudi, A. eds). Praeger, New York, USA.

Munns, D.N. and Mosse, B. 1980. Mineral nutrition of legume crops. Pages 115–125 *in* Forage Legumes for Energy-Efficient Animal Production (Barnes, R.F., Ball, P.R., Brougham, R.W., Marten, G.C. and Minson, D.J., eds) USDA-ARS, USA.

Mytton, R.L., Hughes, D.M. and Kahurananga, J. 1988. Host-*Rhizobium* relationships and their implications for legume breeding. Pages 131–143 *in* Nitrogen Fixation by Legumes in Mediterranean Agriculture (Beck, D.P. and Materon, L.A. eds) Martinus Nijhoff Publishers, The Netherlands.

Norman, M.J.T. 1979. Annual cropping systems in the tropics. University Presses of Florida, Gainesville, Florida, USA.

Norton, G., Bliss, F.A. and Bressani, R. 1985. Biochemical and nutritional attributes of grain legumes. Pages 73–114 *in* Grain legume crops (Summerfield, R.J. and Roberts, E.H. eds). Collins, London, UK.

Nowacki, E. 1980. Heat-stable anti-nutritional factors in leguminous plants. Pages 171–177 *in* Forage Legumes for Energy-Efficient Animal Production (Barnes, R.F., Ball, P.R., Brougham, R.W., Marten, G.C. and Minson, D.J., eds) USDA-ARS, USA.

Papastylianou, I. 1988. The role of legumes in agricultural production in Cyprus. Pages 55–63 *in* Nitrogen Fixation by Legumes in Mediterranean Agriculture (Beck, D.P. and Materon, L.A. eds) Martinus Nijhoff Publishers, The Netherlands.

Pate, J.S., Williams, W. and Farrington, P. 1985. Lupin. Pages 699–746 *in* Grain legume crops (Summerfield, R.J. and Roberts, E.H. eds). Collins, London, UK.

Pauli, A.W., Lechtenberg, V.L., and Wedin, W.F. 1988. Equipment for harvesting, transporting, storing and feeding. Pages 567–594 *in* Alfalfa and Alfalfa Improvement (Hanson, A.A., Barnes, D.K. and Hill, R.R. eds). Agronomy Vol.29. ASA, CSSA, SSA, Madison, Wisconsin, USA.

Polhill, R.M. and van der Maesen, L.J.G. 1985. Taxonomy of grain legumes. Pages 3–36 *in* Grain legume crops (Summerfield, R.J. and Roberts, E.H. eds) Collins, London, UK.

Puckridge, D.W. and French, R.J. 1983. The annual legume pasture in cereal-ley farming systems in Southern Australia: A review. Agriculture, Ecosystems and Environment 9: 229–267.

Raven, P.H. 1973. The evolution of Mediterranean floras. Pages 213–223 *in* Mediterranean Type Ecosystems: Origin and and Structure (di Castri, F. and Mooney, H.A. eds), Springer-Verlag, New York, USA.

Roquero, C. 1979. The potential productivity of Mediterranean soils. Pages 21–42. *in* Soils in Mediterranean Type Climates and Their Yield Potential. Proceedings 14th Colloquium International Potash Institute Dev Bund, A.G. Bern, Switzerland.

Roughley, R.J. 1980. Environmental and cultural aspects of the management of legumes and *Rhizobium*. Pages 97–103 *in* Forage Legumes for Energy-Efficient Animal Production (Barnes, R.F., Ball, P.R., Brougham, R.W., Marten, G.C. and Minson, D.J., eds) USDA-ARS, USA.

Russell, J.S. 1960. Changes in the long term experimental plots at Hybybolite, South Australia. I. Changes in pH, total nitrogen, organic carbon, and bulk density. Australian Journal of Agricultural Research 11: 902–926.

Saxena, M.C. 1988. Food legumes in the Mediterranean type of environment and ICARDA's efforts in improving their productivity. Pages 11–23 *in* Nitrogen Fixation by Legumes in Mediterranean Agriculture (Beck, D.P. and Materon, L.A. eds) Martinus Nijhoff Publishers, The Netherlands.

Saxena, M.C. and Singh, K.B. (eds). 1987. The Chickpea. CAB International and ICARDA. CAB International, Wallingford, U.K.

Schubert, K.R. and Ryle, G.J.A. 1980. The energy requirements for nitrogen fixation in nodulated legumes. Pages 85–96 *in* Forage Legumes for Energy-Efficient Animal Production (Barnes, R.F., Ball, P.R., Brougham, R.W., Marten, G.C. and Minson, D.J., eds) USDA-ARS, USA.

Simmonds, N.W. (ed.) 1976. Evolution of crop plants. Longman, London, UK and New York, USA.

Smithson, J.B., Thompson, J.A. and Summerfield, R.J. 1985. Chickpea. Pages 312–390 *in* Grain legume crops (Summerfield, R.J. and Roberts, E.H. eds). Collins, London, UK.

Stern, W.R. 1985. Environmental and management limitations of legume-based forage systems in Australia. Pages 101–109 *in* Forage Legumes for Energy-Efficient Animal Production (Barnes, R.F., Ball, P.R., Brougham, R.W., Marten, G.C. and Minson, D.J., eds) USDA-ARS, USA.

Stern, W.R. and Donald, C.M. 1962a. Light relationships in grass-clover swards. Australian Journal of Agricultural Research 13(4): 599–614.

Stern, W.R. and Donald, C.M. 1962b. The influence of leaf area and radiation on the growth of clover in swards. Australian Journal of Agricultural Research 13(4): 615–623.

Summerfield, R.J. and Bunting, A.H. 1980. Advances in legume science. Royal Botanic Gardens, Kew, England.

Tainton, N.M. (ed). 1981. Veld and pasture management in South Africa. Shuter and Shooter, Pietermaritzburg, South Africa.

Teuber, L.R., Levin, R.P., Sweeney, T.C. and Phillips, D.A. 1984. Selection for N concentration and forage yield in alfalfa. Crop Science 24: 553–558.

Thrower, N.J.W. and Bradbury, D.E. 1973. The physiography of the Mediterranean lands with special emphasis on California and Chile. Pages 37–52 *in* Mediterranean Type Ecosystems: Origin and and Structure (di Castri, F. and Mooney, H.A. eds), Springer-Verlag, New York, USA.

Webb, C. and Hawtin, G. (eds). 1981. Lentils. Commonwealth Agricultural Bureaux and ICARDA. CAB International, Wallingford, England.

Webber, G., Matz, N. and Williams, G. 1977. Ley farming in South Australia. Bulletin no.15. Department of Agriculture and Fisheries, Adelaide, South Australia.

White, D.H., Elliot, B.R., Sharkey, K.J. and Reeves, T.G. 1978. Efficiency of land use systems involving crops and pastures. Journal of Australian Institute of Agricultural Science 44: 21–27.

Witty, J.F. and Minchin, F.R. 1988. Measurement of nitrogen fixation by the acetylene reduction assay: myth and mysteries. Pages 331–344 *in* Nitrogen Fixation by Legumes in Mediterranean Agriculture (Beck, D.P. and Materon, L.A. eds) Martinus Nijhoff Publishers, The Netherlands.

The Role of Legumes in the Farming Systems of Algeria

A. BENBELKACEM

ITGC, Station Experimentale d'El Khroub, Constantine 25100, Algeria

Abstract. The area grown to food legumes had increased to 250 000 ha by 1987, but yields are low and production does not meet demand, so imports are rising sharply. There is also a deficit in forage legumes. These tend not to be considered as important a source of livestock feed as rangeland, perennial pastures, fallow and cereal residues. However, in 1987 the area for forage crops reached 933 000 ha, rising to more than 1 000 000 in 1988. The main problem with expanding production of food legumes is the lack of varieties adapted to moderate rainfall, resistant to disease, and suitable for mechanization. Other constraints include poor choice of land, inadequate cultural practices and low market prices until 1984. Problems for pastures include degradation of the steppe, poor natural pastures due to bad management, and inadequate crop-livestock integration in the cereal zones. Future research will concentrate in the long-term on breeding, and in the short-term on improvement of management and extension

Introduction

Despite efforts to dynamize agriculture in general, there is a serious deficit in Algeria in the production of cereal grains, food legumes and forages. The national development programs have concentrated more on cereals and made the deficit more acute for forage and food legumes. Investments in machinery, fertilizers, and the establishment of credit and price policies have not been enough to boost production.

Forages

Reliance on the steppe (rangelands), fallow land, permanent pastures and cereal residues (straw and stubble) as a primary cheap food source has downgraded the importance of cultivated forages as a basis for more efficient animal production (cattle, sheep, and goats). The relative importance of the different feed sources over time is shown in Table 1.

In the last six years the area of cultivated forages has more than doubled: it reached 933 thousand hectares in 1987 and will be over one million hectares in 1988. Livestock numbers are increasing steadily however (Table 2), and in spite of the doubled area domestic production of feed

A.E. Osman et al. (eds.), The Role of Legumes in the Farming Systems of the Mediterranean Areas, 31–37.

Table 1. Area of the different feed resources (1000 ha) and their percentage of total feed area 1973–81.

Year	Cultivated forages Area	%	Perennial pastures Area	%	Cut fallow Area	%	Grazed fallow Area	%	Total feed area
1973	187	11.0	40	2.4	105	6.2	1366	80.4	1698
1975	239	10.8	24	1.1	186	8.4	1767	79.7	2216
1977	270	12.6	20	0.9	101	4.7	1751	81.7	2142
1979	349	17.1	15	0.7	94	4.6	1583	76.6	2041
1981	415	20.3	17	0.8	70	3.4	1541	75.4	2043

Source: Maatougui (1987); ITGC (1987).

Table 2. Livestock numbers (in 1000) 1973–81.

Year	Cattle	Sheep	Goats
1973	872	8455	2406
1975	1002	9772	2269
1977	1130	10298	2421
1979	1327	12222	2817
1981	1376	13739	2749

Source: Maatougui (1987); ITGC (1987).

cannot meet the demand. The National Agricultural Institute at El Harrach sees the need for more research into animal feed requirements (Arab Agriculture 1988).

Between 1973 and 1981 annual growth rates were 1.55% for cattle and 3.62% for sheep, whereas for goats it tended to decrease, at 1.11%.

Cultivated Forage Crops

Forage crops were grown on average at the following proportions of the total cultivated forage area 1973–81:

Vetch + oat	70.0%
Barley + oat + rye	8.3%
Berseem	2.1%
Alfalfa	3.4%
Sorghum and maize	1.4%
Forage beet	0.3%
Others	14.5%

Hay forages are clearly the most important, and are mostly used for sheep. Berseem, alfalfa, sorghum, maize, and beet are used as green fodder and directed more towards cattle. Barley, oats and rye are dual-purpose crops for both grazing and haymaking. 'Others' mainly represent annual medics which was widely introduced as a replacement for grazed fallow.

There is a noticeable absence of many other possible forage crops. In 1981 medics were discarded and the forage pea was introduced as a partial replacement for vetch and other crops such as rye grass. Faba bean as grain fodder and triticales were also introduced on a smaller scale. Annual medics were reintroduced in 1986.

Food Legumes

Food legumes have been cultivated since ancient times for their grain production and protein content. They are used to balance the deficits in animal proteins in the developing countries. As with forages, the production deficit in food legumes in Algeria is still acute. In the past they were grown on a large scale, mainly to meet the needs of the French, in Algeria and in France. In 1925, for example, more than 40,000 t were exported, and in 1953 over a hundred thousand hectares were cultivated with food legumes. When the French left in 1952 much of the land was abandoned, management deteriorated and the area and productivity of food legumes decreased: in 1963 the area was down to fifty thousand hectares. The population was increasing, however, and people were eating more legumes. Also, yields were poor. Imports began to increase, from 2.4 million tonnes in 1974 to 9.8 million tonnes in 1987, and gradually more and more land was brought back into food legume cultivation. In 1986–87 250 thousand hectares were sown to food legumes, a four-fold increase (ITGC 1982).

A yearly variation in yield levels is evident from Table 3, implying poor adaptation of varieties to climatic fluctuations. Also it will be observed that lentil yields are far lower than those of other crops.

Table 3. Food legume yields (kg/ha).

Year	1964	1970	1975	1980	1985
Crops:					
Faba bean	920	590	970	540	750
Beans	740	230	970	370	570
Chickpea	510	450	730	390	520
Peas	600	340	630	310	470
Lentil	370	290	470	180	330

Source: ITGC (1982–1986/87); ITGC (1982).

Present Role of Legumes

The actual needs of the population were estimated at 7 kg/year per person, so it is possible to work out the area that is needed to grow enough to satisfy these needs, based on yields obtained from experiments using optimal varieties and cultural techniques (Table 4). For example, experimental yield levels reached an average of 2000 kg/ha for lentils.

Table 4. Area (1000 ha), yield (kg/ha), and production (t) estimated for each crop to meet national needs.

Crops	Area	Yield	Production (t)	% need satisfaction
Faba bean	70	1000	70000	182
Chickpea	70	800	56000	109
Lentil	30	700	21000	102
Pea	20	700	14000	107
Beans*	10	450	4500	10

* The possibility for extending this crop is limited because varieties adapted to moderate rainfall are not available.

Source: ITGC (1982).

Physical Environment

Algeria covers a very large area, over two million km^2, but a vast proportion of this consists of arid plateaux, mountains and desert. There are approximately 7.5 million hectares of cultivable land, in the narrow coastal strip. The other two regions are the mountain ranges; Tel Atlas, High Plateaux and Sahara Atlas; and to the south, the Sahara Desert.

Climate

In the north winters are warm and wet, with average rainfall reaching 1000 mm in places, and summers are hot and dry. Further south winters are colder, and summer droughts can last six months. Rainfall is between 200 and 400 mm but is very unpredictable in timing and quantity. The Sirocco, known locally as the 'Chehili', a scorching, dry, dusty wind from the south, blows for more than a month in the Plateaux, and rather less nearer the coast. In the desert, temperatures average about 30°C, (55°C has been recorded), while rainfall is less than 130 mm, and extremely irregular; several cm may fall in one day, and then it may not rain at all for several years (The Middle East and North Africa 1984).

Mechanization

The level of mechanization has not kept pace with agricultural growth. In 1985 there was one tractor/154 ha and one combine/730 ha of cultivated land. However, the Office National de Material Agricole (ONAMA) is aiming at one tractor/73 ha, and one combine/300 ha in the near future. There are factories making all kinds of agricultural machinery in Constantine, Sidi bel Abbes and Berronaghia, and the ultimate goal is to produce every piece of equipment locally. To promote research and development new technology will be developed and tested on about 120 pilot farms. Extension service technicians will train private sector farmers in the use of new technologies and equipment.

Socio-economic Considerations

The 1985–89 development plan is giving agriculture 'absolute priority', and the government is trying to make it more efficient by providing credit facilities, redistributing state land to private farmers, providing chemicals and machinery, relaxing price controls and liberalising marketing. It is also trying to prevent the urban drift by improving rural conditions, for example, providing better housing and electricity.

Constraints to Legume Production

An analysis of the pasture forage situation points to the following problems:

1. Declining resources in the steppe through overgrazing, wind erosion and desertification.
2. Poor regeneration of pastures because of overgrazing, resulting in low vegetation density with low nutritive value.
3. Inadequate crop-livestock integration in the cereal zones, neglecting the important potential of grazed fallow (over one millon hectares) and contributing to degradation of the steppe.
4. Serious weed problems generated by grazed fallow and affecting cereal yields.

The most serious problem limiting the expansion of food legumes is the lack of suitable cultivars that would allow higher yields. Planting of chickpea in winter is impossible because of the high risk of disease, especially ascochyta blight. The crops are therefore grown in the spring with high risk of drought and poor growth, making mechanical harvesting difficult. This is particularly true for lentils, which are more affected by shattering. The local varieties have been used for a long time now and are poor in yield and disease resistance.

Other constraints to increasing food legume production are:

1. Poor choice of land for growing food legumes
2. Unadapted cultural practices, such as soil preparation, seed bed preparation and weed control.
3. Losses due to hand-harvesting.
4. Lack of machinery.
5. Low prices of products until 1984 (Fig. 1).

Future Strategies

The 1985–89 development plan aims to develop agriculture generally by increasing the area of cultivated land, and making more intensive use of it. Plans include developing new land, especially the steppe, reclaiming fallow, introducing integrated farming in mountainous and desert regions, and extending irrigated areas. The least productive vineyards are to be con-

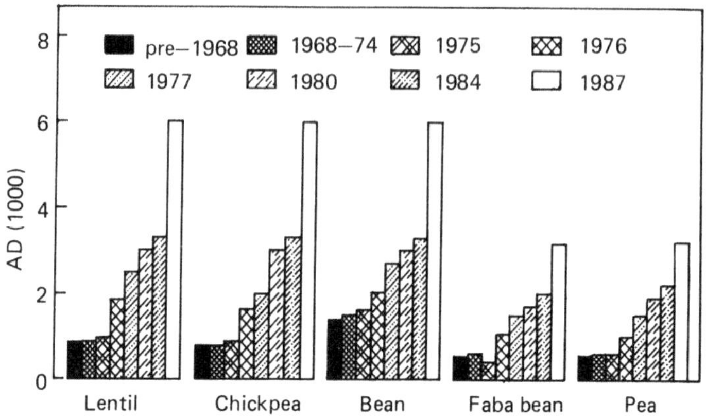

Source of data:: ITGC (1982); Maatougui (1987).

*AD-- Algerian Dinars (1USD – 7AD approx.)

Fig. 1. Grain legume prices (AD*/t) pre-1968 to 1987.

verted to cereal and dairy farming. The government is also investing in factories to produce enough phosphate and nitrogenous fertilizers for the country's own needs and an exportable surplus.

In particular, one objective is to increase pasture and grain legume production. Where there is more than 500 mm annual rainfall highly productive pastures can be developed with annual and perennial legumes (*Trifolium* spp, *Medicago* spp, *Hedysarum* spp) and annual and perennial grasses (*Lolium* spp, *Dactylis* spp, *Phalaris* spp). In the areas receiving 300-500 mm, there is great potential for developing annual pastures as a replacement for fallow with annual medics and subterranean clovers, and more integration of forage pea, forage triticale and non-irrigated alfalfa.

In order to solve the problems of food legumes there are two main strategies:

1. Long-term, through breeding, to make available reliably high-yielding cultivars resistant to diseases and of suitable height for mechanised harvesting (>30 cm). There has been a very successful program of exchange of germplasm mainly with ICARDA, but also with ICRISAT, which has given rapid results through material selection and adaptation that will surely improve the situation of food legumes at farmer level, especially for lentil and chickpea, where many lines emerged. These are now being multiplied (ILC 3279, 482 and 195; NEC 105; NEL 285,468 and 780).
2. Short term, through a good package of cultural practices and a larger extension service network, to improve management standards. The reinforcement of contacts between farmers and researchers is emphasized as an important and necessary task.

The ITGC (Institut Technique des Grandes Cultures) is directing the research and extension efforts, but much more remains to be done. The stimulation of the actual level of production by improving the use of all Algeria's resources, both material and human, should be at the roots of the strategy to raise agricultural productivity.

References

ITGC, 1982, Seminaire regional sur le developpement des legumes secs en Algerie, Constantine, Algerie.
ITGC: Bilans annuels des campagnes 1982 a 1986–87.
Revue Cerealiculture No.16, ITGC, 1987.
Maatougui, M.E.H. Country report submitted for the 1987 ICARDA International Workshop on Genetic resources of cool season pasture, forage and food legumes for semi-arid temperate environments, June 1987, Cairo, Egypt.
Arab Agriculture 1988 Yearbook, Arab World Agribusiness, Falcon Publishing WLL, Bahrain.
The Middle East and North Africa 1984–85 Europa Publications 1984, London, UK.

Discussion

Gintzburger:
With reference to your statement that grain legumes were usually cultivated on rough and marginal land, has the increase in the price of these legumes since 1984 changed the areas where they are grown?

Benbelkacem:
All the grain legumes are grown in the sublittoral and high plateaux regions. Since the price increase they will continue to be grown there, but more intensively. Food legumes are grown in fallow now.

The Role of Legumes in the Farming Systems of Cyprus

I. PAPASTYLIANOU

Agricultural Research Institute, Nicosia, Cyprus

Abstract. The present report presents experimental evidence demonstrating the potential importance of legumes in the rainfed areas of Cyprus. Legumes fixed up to 80% of their total nitrogen production from the atmosphere. Barley following legumes gave higher yields than barley after cereals. The mean annual nitrogen balance (output minus input) in a vetch-barley rotation was 60 kg N/ha, while in a continuous barley rotation it was only 6 kg N/ha. Though the contribution that legumes can make is well demonstrated, their cultivation is still limited. The constraints on expansion of cultivation, such as unfavourable economics and government policy, low productivity, and difficulties in mechanization, are discussed.

Introduction

The group of plants belonging to the Leguminosae are known agronomically for two main characteristics: (a) their symbiosis with *Rhizobium* bacteria, which fix nitrogen (N) from the atmosphere, and (b) the high protein concentration of their products. Because they can obtain N from the atmosphere, legumes can be independent of fertilizer N, reducing the cost of production and minimizing pollution of underground water. Because of the high N concentration of their tissues, legumes can increase soil N for other crops. In addition, having legumes in rotation with other crops is a management technique which minimizes weeds, pests and pathogens for the subsequent crop.

The high protein concentration of leguminous products makes them very valuable food for humans and animals. In countries where the price of meat is too high for most people, legumes are the best protein substitute. In developed countries, legumes are the basic feed for animals in the process of animal protein production.

In Cyprus, legumes are cultivated on only 10–15% of the rainfed areas, the rest of the area being sown to cereals. The predominant farming system in these areas is continuous barley production. The area sown to legumes and the production of legumes has not increased in the last ten years. Approximately 80% of the protein requirements for animal nutrition is imported, while animals' needs for roughage are met by hay and straw produced in Cyprus. However, small grain cereals cover only 15% of the

A.E. Osman et al. (eds.), The Role of Legumes in the Farming Systems of the Mediterranean Areas, 39–49.

country's requirements for animal feedstuffs. It is worth noting that there is no local production of protein supplement feedstuffs (T. Antoniou, personal communication).

Though the contribution of legumes is well known, their cultivation in West Asia and North Africa (WANA) is still limited. This paper will give experimental evidence demonstrating the high potential of legumes in Cyprus, and discuss the constraints limiting the expansion of legumes in the region.

Physical Environment

Soils in areas where rainfed crops are grown in Cyprus are calcareous, of medium to fine texture, with low nitrogen (total soil nitrogen around 0.1%) and phosphorus (bicarbonate-extractable 1–5 ppm), but adequate potassium (exchangeable K >250 ppm).

Rainfall occurs during November through April and is rather erratic, both in amount and distribution: 40% of the rainfall occurs in December and January. In the cereal-growing areas, average annual rainfall is around 300 mm. Water stress may occur at any time in the growth cycle, but the crop is most likely to suffer in March-April during anthesis and grain-filling, reducing grain yield; and during sowing, resulting in late emergence and thin stands.

Present Role of Legumes in Farming Systems

Nitrogen Fixation

The amount and percentage of nitrogen fixed by legumes commonly grown in Cyprus are presented in Table 1. These values were estimated using ^{15}N methodology in a cooperative project with the International Atomic Energy Agency (IAEA). In these studies there was no inoculation so the legumes depended on indigenous populations of *Rhizobium* for nitrogen fixation. Rainfall at the experimental site (Dromolaxia) in the two years of the study was 234 and 330 mm. Low values of nitrogen fixation (around 60%) were obtained in 1982/83 while in the next season, values were generally higher (70–80%). Of the legumes studied, only chickpea had low nitrogen fixation, due to poor nodulation.

Most of the rainfed legumes obtained 60–80% of their total nitrogen from fixation. This is considered satisfactory under the prevailing conditions, which included moisture stress during the first year and no precautions for pathogen control. *Sitona* weevil, which damages nodules, has been frequently observed in Cyprus, especially on faba beans and common vetch. However, the extent to which nitrogen fixation is reduced by this weevil has not yet been determined under our conditions.

N_2 fixation by legumes should not be taken for granted, however. It can

Table 1. Nitrogen fixed (kg N/ha and %N derived from fixation) by legumes in two seasons in Cyprus, using the A-value method, with oats (1982/83) and barley (1983/84) as reference crops.

Crops	1982/83		1983/84	
	kg N/ha	%df*	kg N/ha	%df
Common vetch (*Vicia sativa* L.)**	93	73	106	75
Peas (*Pisum sativum* L.)**	23	36	91	63
Medics (*Medicago truncatula* Gaerth)				
cv Jemalong (1982/83)	122	67		
cv Cyprus (1983/84)			90	70
Ochrus vetch (*Lathyrus ochrus* (L.)				
A.P. de Candole)	71	66	105	79
Bitter vetch (*Vicia ervilia*				
(L.) Willd.)	45	56	105	70
Broad beans (*Vicia faba major*)	122	63	176	80
Tick beans (*Vicia faba minor*)			180	84
Woollypod vetch (*Vicia dasycarpa* Ten)			151	79
Chickpea (*Cicer arietinum* L.)			25	41

* % of total plant nitrogen derived from fixation.
** Oat was reference crop in both seasons.

Source: Papastylianou, 1988a.

take place only if an appropriate strain of *Rhizobium* is present in the soil. As shown in Tables 1 and 2, an appreciable amount of nitrogen is fixed by many legume species in Cyprus, indicating the effectiveness of the nodules present on these crops. The possibility of improving the indigenous popula-

Table 2. Percentage of N_2 fixation by vetch (V) and pea (P) grown in pure stands and in mixtures with oats (OV_1, OV_2, OP_1 and OP_2) at two levels of N fertilization (15 and 90 kg N/ha) for two seasons (1982/83 and 1983/84).

Crop	1982/83			1983/84		
	15 N	90 N	Mean	15 N	90 N	Mean
V	73.3	38.3	55.8ab	74.7	47.5	61.1a
OV_1	59.4	27.6	43.5	66.0	56.2	61.1a
OV_2	74.8	56.7	65.7a	78.5	48.2	63.4a
P	35.8	18.5	27.2c	63.5	29.2	45.8ab
OP_1	57.8	45.7	51.7ab	47.7	31.7	39.7b
OP_2	56.9	40.8	48.8b	64.5	18.5	41.5b
Mean	59.7	37.9		65.8	38.5	

- S.E. for comparing the means of the fertilizer x crop for the 1983/84 season:4.81.
- The crop x fertilizer interaction in 1982/83 was not significant.
- The difference between the two fertilizer means was significant in 1982/83 but not significant in 1983/84.
- Means in a column followed by the same letter n.s. at 5% by Duncan's new multiple range test.

Source: Papastylianou, 1988b.

tion of *Rhizobium* by selection and introduction is under consideration. However, the need for inoculation is urgent for a few legume species in Cyprus either because no nodules form, or those present do not have the characteristic pink color which indicates effectiveness. Crops with such problems are chickpea under rainfed conditions, soybean, peanut and field bean.

Nitrogen fertilization or high levels of soil mineral N suppress N_2 fixation. This was shown in an experiment in which vetch and field peas were grown in pure stands and in mixtures with oat (Table 2).

Most of the legumes grown by farmers under rainfed conditions are included in Table 1. Medics are not cultivated crops but they grow naturally on uncultivated land. With the exception of chickpea, cultivated legumes of the genera *Lathyrus*, *Vicia* and *Pisum* respond to *R.leguminosarum* biovar. *viceae*. For crops which respond to *Rhizobium* species other than *R. meliloti* and *R. leguminosarum*, it seems that there are nodulation problems.

The results of the first inoculation studies are presented in Table 3. Inocula were supplied by Nitragin Company (USA) and ICARDA. The data show that an appreciable yield response could be obtained by inoculating non-nodulated legumes such as soybean, peanut and chickpea. Vetch, which can fix 70% of its nitrogen (Table 1), also showed a yield response with inoculation. Inoculation studies of field bean (*Phaseolus*) have not yet shown any response due to the presence of high mineral nitrogen in the irrigated areas where the experiments were carried out.

Although vetch showed a response to inoculation (Table 3), it did not respond to nitrogen fertilization. In studies on vetch carried out over many years and locations in Cyprus, dry matter and nitrogen yields were not increased with nitrogen fertilization(Table 4). In contrast, nitrogen fertilization had a negative effect on nitrogen fixation. In Dromolaxia in 1983 and 1984 the amount of nitrogen fixed by vetch was reduced by 48% and 36% respectively when 90 kg N/ha was applied.

Table 3. Effect of inoculation on dry matter and nitrogen yield of four crops in Cyprus.

Crop	Yield (kg/ha)[1]		N Yield (kg N/ha)	
	Inoculated	Uninoculated	Inoculated	Uninoculated
Soybean	4853(+)	3677(−)	302(+)	212(−)
Vetch	5020(+)	4366(+)	120(+)	107(+)
Chickpea	2628(+)	1683(−)	71(+)	41(−)
Peanut	4279(+)[2]	2967(−)	123(+)[3]	58(−)

[1] Grain yield for soybean and dry matter yield for vetch and chickpea; (+) and (−) denote the presence or absence of nodules respectively.
[2] Yield of pods.
[3] Yield of N from whole plants excluding pods.

Source for soybean data: Papastylianou 1986.

Table 4. Performance of vetch grown in Cyprus at different locations and years at two levels of nitrogen fertilizer.

Year	Location	Fertilizer (kg N/ha)	DM yield (kg/ha)	Nitrogen yield (kg N/ha)
1981/82	Laxia	0	4090	112
		60	4320	117
1982/83	Kivisili	0	4609	112
		60	4553	104
1983/84	Athalassa	0	5079	
		60	4305	
1982/83	Dromolaxia	15	4714	127
		90	3105	100
1983/84	Dromolaxia	0	7004	165
		90	7100	171

Source: Papastylianou 1988a.

In studies with soybean, nitrogen fertilization and inoculation gave equal grain yields. Both treatments yielded 25% higher than the uninoculated, unfertilized control.

Legumes in Rotation With Cereals

The residual effect of legumes on subsequent crops and the protein yield, which is important for human and animal nutrition, were studied in rotation experiments carried out during four growing seasons at two rainfed sites in Cyprus. In Table 5, grain yield of barley grown after vetch is compared with

Table 5. Grain yield (kg/ha) of barley grown in vetch-barley or continuous barley for grain rotations with different levels of nitrogen fertilization at two locations, 1981–85.

Rotation	Fertilizer on barley (kg N-ha)	Laxia				Dromolaxia			Mean
		1981/82	1982/83	1983/84	1984/85	1982/83	1983/84	1984/85	
	0	1232	664	1434	2550	1187	1030	3014	1586
Vetch*-Barley	30	2188	1016	1367	2237	1426	1277	3476	1855
	60	2121	856	807	1019	1680	1419	3807	1572
	90	2965	829	672	684	1524	1464	3901	1719
	0	239	560	336	736	545	396	2643	779
Barley-Barley	30	986	739	1113	1748	769	537	2922	1259
	60	1845	1299	1269	1569	948	1277	2601	1544
	90	1762	829	634	768	1113	1000	3115	1317
	120	2069	889	358	391	1143	929	3029	962
Rainfall mm	226	207	255	265	234	330	487	286	

* Vetch was not fertilized.

Source: Papastylianou 1988a.

grain yield in a continuous barley growing system. The data show that (a) barley after vetch without any nitrogen fertilization yielded as much as continuously cropped barley supplied with 60 kg N/ha, (b) barley fertilized with unlimited nitrogen in continuous cropping cannot yield as much as barley after vetch when it is fertilized with 30 kg N/ha, and (c) high doses of nitrogen fertilizer (>60 kg N/ha) cause yield reduction in both rotations.

The total nitrogen yields of two rotations are presented in Fig. 1. The rotations were vetch-barley for grain without nitrogen fertilization, or barley for hay followed by barley for grain with 60 kg N/ha applied in both phases. These treatments were chosen to demonstrate the importance of legumes in rotations in countries where farmers do not apply any nitrogen fertilizer.

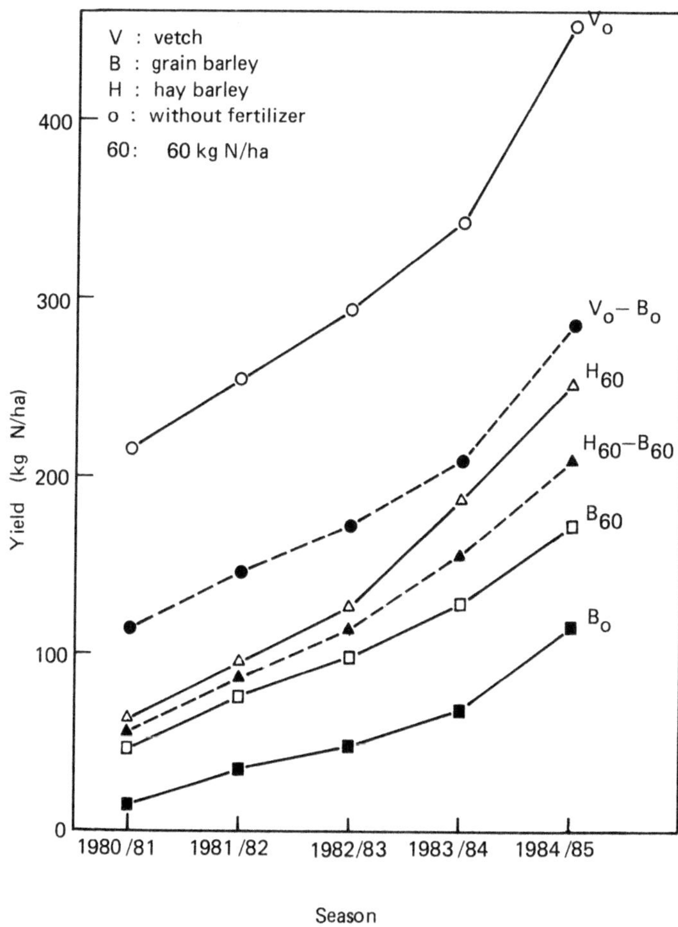

Fig. 1. Cumulative nitrogen yield of crops in vetch-grain barley rotation without fertilizer and in hay barley-grain barley rotation with 60 kgN/ha in both phases 1980–1985, Laxia, Cyprus. (Papastylianou 1988a).

Undoubtedly, nitrogen output from crops in the rotation which included legumes is higher than in the hay barley-grain barley rotation.

Considering the results in Table 5 and Fig. 1, the best rotation practice should be vetch (without nitrogen fertilization) – barley for grain supplied with 30 kg N/ha, for conditions similar to the experimental sites. These results and conclusions will be tested with further cycles of these rotation studies.

The N economy of rotations including legumes compared to monoculture of cereals is shown in Table 6. The N balance (output minus input) over six years for a rotation with barley as the forage crop was around + 35 kg N/ha, while that for a rotation with woollypod vetch as forage was around + 300 kg N/ha.

In short-term rotation studies, the residual effect of a number of legumes on the performance of subsequent barley was also demonstrated (Table 7).

Table 6. Output of N of barley for grain (B) and forages (Woollypod vetch (L) and barley (F)), input of N fertilizer and balance of N in the rotations for the period 1977–82.

Rotation	Nitrogen removed in barley grain plus straw		Nitrogen removed in the forage	Input of N fertilizer	Balance of N (output-input)
LB	195		292	130	357
L*B*B	122*	241	195	174	262
LB*B*	119*				
FB	157		134	261	30
F*B*B	105*	210	89	261	38
FB*B*	105*				

* Data presented are for the crop *in italics*.

Source: Papastylianou and Samios 1987.

Table 7. Grain (t/ha), nitrogen yield (kg/ha) and nitrogen concentration (%) of barley as affected by the preceding crops in two separate cycles.

Preceding crop	Growing season 1983/84			Growing season 1984/85		
	Grain yield	Nitrogen yield	Nitrogen concentration	Grain yield	Nitrogen yield	Nitrogen concentration
Barley	1.72	28.1	1.63	1.58	24.9	1.58
Oat	1.80	29.5	1.65			
Ryegrass				1.04	16.5	1.57
Medic	2.08	42.1	2.05	3.92	63.8	1.62
Ochrus vetch	2.54	45.7	1.81	3.52	54.6	1.55
Bitter vetch	2.38	44.6	1.86	3.49	71.0	1.57
Faba bean	2.32	41.3	1.77	4.72	79.7	1.66
Tick bean				3.49	71.0	1.57
Woollypod vetch				4.84	83.6	1.72
Chickpea				2.91	45.5	1.54
S.E.	0.133	2.88	0.076	0.378	7.55	0.055

Source: Papastylianou 1987b.

Research

Constraints on the Expansion of the Cultivation of Legumes

The contribution of legumes in terms of the N economy of the soil and in their residual effect on subsequently-grown cereals is well-demonstrated, but there are other factors at play, which limit their cultivation. In the next part of this report a farming system including legume cultivation is compared to the farming systems which it would replace, i.e. fallow-barley or continuous barley: likely problems are identified and possible solutions discussed.

Mechanization

Machines can be adjusted and modified to handle the sowing and harvesting of most legumes. In addition, crop varieties with characteristics which facilitate machine-harvesting are available: for example, determinant types of faba bean bearing pods at the top of the plant, and the tall chickpea (Hadjichristodolou 1984). Problems arising from the shattering of pods of forage legumes, and the harvesting of peanuts and *Phaseolus* bean have been solved easily by adapting existing machinery. The main constraint to mechanization is the cost of machines. This could be solved by groups of farmers buying machinery on a cooperative basis.

Weeds

Weeding of legumes is mainly done by hand in most of the WANA region. Weedicides are now available for most of the legume crops, but the cost of these chemicals is a limiting factor. However, this should be considered in relation to the general economics of production: it could be feasible if productivity and income are increased by new varieties and better management.

Productivity

Low yields are one of the main limiting factors for legume cultivation. Most of the winter-grown legumes need higher temperatures and higher rainfall than barley. Cold- and drought-resistant varieties are desirable for most of the winter legumes. This is not an impossible target, as the chickpea variety ILC 3270 has shown.

Cultural practices, such as inoculation, P and K fertilization, correct plant population, and control of weeds, pests and pathogens should be applied to allow legumes to maximize their production. In many cases when nodules do not form, legumes are not fertilized with N. Legumes must either be inoculated with *Rhizobium*, which is the correct agronomic and economic practice, or must be N fertilized.

As already mentioned, legume breeding should have yield increase as one target. If for a particular legume the same breeding effort were to be made as was made for barley, the same chances exist for yield increase. This point must be drawn to the attention of plant breeders and the different institutions as they plan their future programs, if legumes are to be properly promoted.

Plant Breeding

Plant breeding is usually a means of solving problems, but it is considered here as a problem itself because as far as legumes are concerned it is not used properly or to a sufficient extent. Many of the problems of growing legumes such as cold susceptibility, unsuitability for mechanization, low productivity and susceptibility to diseases and pests could be solved by new varieties. As a result of the most basic program of germplasm introduction and screening, the chickpea variety ILC 3279, which is tall for ease of mechanical harvesting, has cold and ascochyta resistance and is more productive than the traditional varieties, is now widely grown in many countries of the WANA region. Similar breeding efforts can solve problems more cost-effectively than the use of chemicals, and the cost of production will drop with more productive varieties.

Marketing

For cereals, there is a well organized world and, in many countries, local market. For legumes such a market exists only for soybean. Most of the grain legumes are consumed by humans and generally the possibility of increasing local consumption is rather limited. Varieties which are more productive and products which can be used as animal feedstuffs will expand the market for these legumes. International marketing of hay produced by forage legumes is even more difficult, because of transportation cost. However, in countries where there is a shortage of roughage, marketing of forages will not be a problem because hay can replace imported concentrates.

Government Policy

Subsidies, crop insurances, and marketing of agricultural products are facets of government policy which can encourage or discourage the cultivation of a crop and determine the whole farming system of a country. In Cyprus, cereals are covered by insurances, are heavily subsidized and have a well organized market. Such a situation does not exist yet for any legume. Farmers grow what is economically safe and not what is agronomically correct.

Conclusion

This report has highlighted the fact that legumes can make valuable contributions to agricultural production in Cyprus and has touched very briefly on the different constraints to the expansion of legume cultivation in the WANA region. General suggestions have been given in an attempt to stimulate organized research on this topic.

It can be concluded that the two main factors limiting the expansion of legumes are productivity and economics. To increase productivity is a task for the scientist, while the economics depend to a great extent on government policy. The introduction of legumes to the farming systems of countries of the WANA region should be a joint enterprise for scientists of many disciplines and for governments.

References

Hadjichristodoulou, A. 1984. New chickpea varieties for winter sowing and mechanical harvesting. Technical Bulletin 5B. Agricultural Research Institute, Nicosia, Cyprus. 8pp.

Papastylianou, I. 1986. Effect of nitrogen fertilization and inoculation with rhizobia on nodulation and nitrogen and grain yield of soybean. Miscellaneous Reports No 26. Agricultural Research Institute, Nicosia, Cyprus. 6pp.

Papastylianou, I. 1987a. Amount of nitrogen fixed by forage, pasture and grain legumes in Cyprus, estimated by the A-value and a modified difference method. Plant and Soil 104: 23–29.

Papastylianou, I. 1987b. Effect of preceding legume or cereal on barley grain and nitrogen yield. Journal of Agricultural Science, Cambridge 108: 623–626.

Papastylianou, I. 1988a. The role of legumes in agricultural production in Cyprus. Pages 55–63 *in* Nitrogen fixation by legumes in Mediterranean agriculture (Beck, D.P. and Materon, L.A. eds). Martinus Nijhoff Publishers, The Hague, The Netherlands.

Papastylianou, I. 1988b. The N-15 methodology in estimating N_2 fixation by vetch and pea grown in pure stand or in mixtures with oat. Plant and Soil 107: 183–188.

Papastylianou, I and Samios, Th. 1987. Comparison of rotations in which barley for grain follows woollypod vetch or forage barley. Journal of Agricultural Science, Cambridge 108: 609–615.

Discussion

Osman:
Your data show that legumes in the rotation contribute something more than fixing nitrogen. Could that be due to control of root diseases?

Papastylianou:
We measured the severity of leaf diseases this season and found that barley-after-vetch and -fallow had less damage than after-barley. The measurements were taken in long-term rotation studies where the differences in the effect of rotations on leaf disease control was visually obvious. Soil-borne diseases have not been studied yet in these experiments.

Buddenhagen:
Why does the government support barley at such an exorbitant price above the world market?

Papastylianou:
To keep the farmer in business. If the subsidies are withdrawn the farmers will turn to more profitable business, like the tourist industry. Farmers' unions are pushing for the subsidy.

Durutan:
What is the rate of P_2O_5 applied to barley and vetch?

Papastylianou:
The same amount to both, 20kg/ha.

The Role of Legumes in the Farming Systems of Egypt

A.M. NASSIB, A. RAMMAH and A.H.A. HUSSEIN

Field Crops Research Institute, ARC, Giza 12619, Egypt

Abstract. The most important legume crop in Egypt, growing on almost half of the winter cropping area, is berseem (*Trifolium alexandrinum*), a forage crop. The food legumes, faba beans, chickpea, lentil and pea, are considerably less important although acreage, production and yield has been increasing since 1981. Research programs include breeding for better N fixation, resistance to pests, diseases and parasites, and early maturity; and experiments to improve production methods, particularly mechanization. On-farm research has been well demonstrated by the ICARDA/IFAD Nile Valley Project, and has been adopted by the ARC. Research-extension linkages need strengthening.

Introduction

A single winter forage legume, Egyptian clover, or berseem (*Trifolium alexandrinum*) contributed 22.9% of the total value of field crops production in Egypt in 1986. It was only marginally exceeded by the cotton crop. The clover acreage fluctuates annually around 1.18 million hectares. Amongst the other forage crops the winter sown vetch (*Lathyrus sativus*) and fenugreek (*Trigonella foenum graecum*) are grown to a very limited extent and are restricted to South Egypt.

The only perennial forage crop grown is alfalfa (*Medicago sativa*), but its cultivation is prohibited in the Nile Delta and Valley since it hosts the cotton leaf worm. About 5 500 hectares of alfalfa is grown on the old lands in some districts in Aswan and Qena where there is no cotton. Alfalfa has, however, assumed importance as a crop in the development of the newly reclaimed lands particularly in the regions west of the Delta and occupies about 42 000 hectares.

The four food legumes, i.e., faba bean, pea, chickpea and lentil contributed 3.46, 1.25, 0.24 and 0.30%, respectively to the total production value of field crops in 1986. The acreage of the four legumes increased during 1979–1987, though the increases in pea and chickpea were more prominent (Figs. 1, 2, 3 and 4). The average yield of the four legumes increased during the same period (Fig. 5) and stood at 2.8 t/ha for faba bean, 1.83 t/ha for dry pea, 1.61 t/ha for chickpea and 1.71 t/ha for lentil in 1987; the mean

A.E. Osman et al. (eds.), The Role of Legumes in the Farming Systems of the Mediterranean Areas, 51–61.

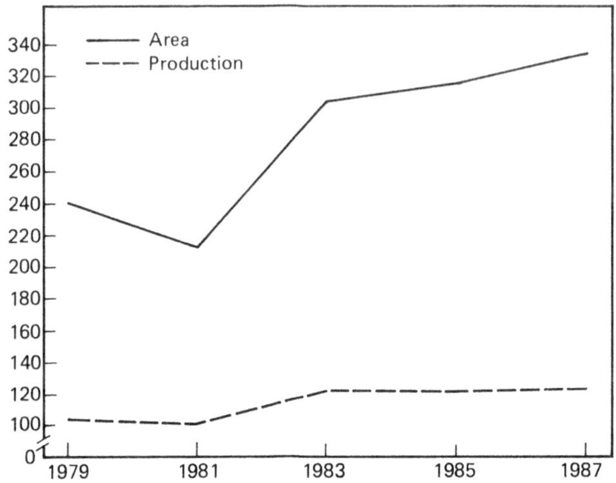

Fig. 1. Faba bean: area (1000 ha) and production (1000 t).

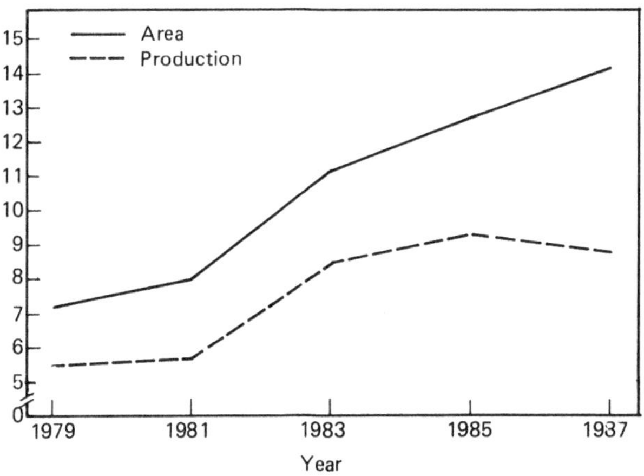

Fig. 2. Pea: area (1000 ha) and production (1000 t).

acreage for the above crops in the same years was 113,532, 7700, 7500 and 5000 ha respectively.

The total production of faba bean and lentil increased substantially during 1981–1987 (Figs. 1 and 4). This is due to increasing farmer interest in both crops because they fetch attractive prices in the free market. In addition, some lentil production moved from South Egypt to non-traditional areas in the Nile Delta which appear to be agronomically more favorable. The

Fig. 3. Chickpea: area (1000 ha) and production (1000 t).

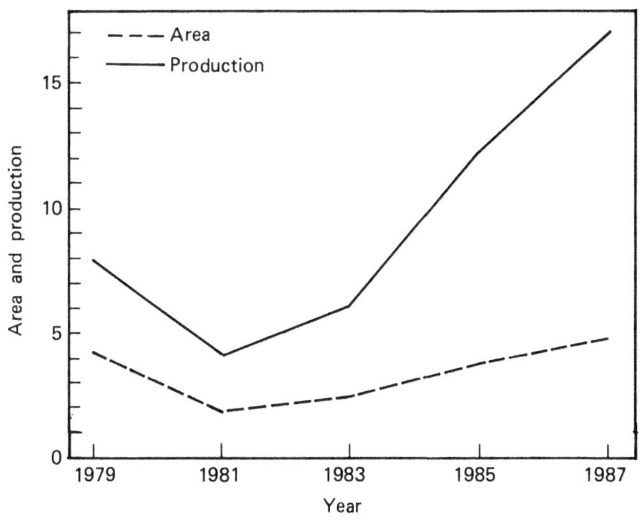

Fig. 4. Lentil: area (1000 ha) and production (1000 t).

national self-sufficiency rate increased from 68% to 87% for faba beans and from 7% to 35% for lentil during 1980–1985 (Table 1). The per caput consumption for both crops has varied greatly during 1970–1985. The main factor seems to be availability of faba bean (Watson 1981) and this could apply also to lentil. Imports of lentil have declined since 1984, and of faba bean stopped altogether in 1983. The country is self-sufficient in pea and chickpea.

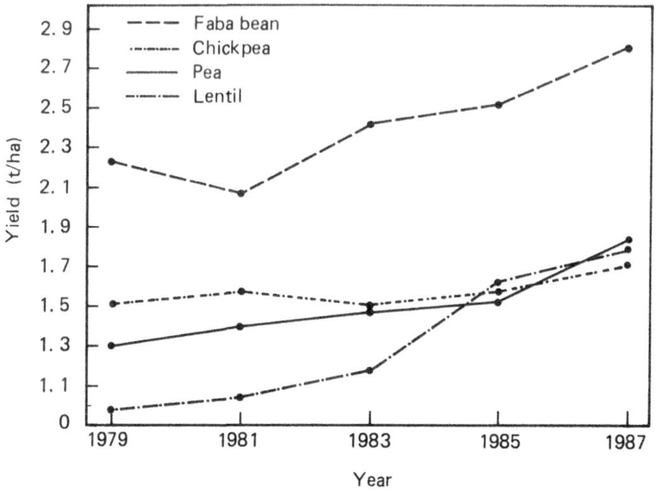

Fig. 5. Average yield (t/ha) of faba bean, pea, chickpea and lentil, 1979–87.

Table 1. National (1000 t) and per caput (kg/year) consumption, imports (1000 t), and self-sufficiency rate (%) of dry faba bean and lentil 1970–87.

Year	National consumption	Per caput consumption	Imports	Self-sufficiency
Faba bean				
1970/71	206	6.2	2	115
1975	312	8.7	110	80
1980/81	251	6.0	70	68
1984/85	284	6.2	0	87
1985/86	258	–	0	–
1986/87	228	–	0	–
Lentil				
1970/71	57	1.7	0	88
1975	73	2.0	53	53
1980/81	67	1.6	61	7
1984/85	38	0.8	13	35
1985/86	17	–	18	–
1986/87	12	–	2	–

Physical Environment

Climate

Agriculturally, Egypt may be roughly divided into two climatic regions. The first includes the Delta and is characterized by a Mediterranean-type climate. The winter is mild with maximum temperature of 20°C and minimum temperature of 7°C. Winter rainfall ranges from 60 mm (east) to

200 mm (west) in the northern coastal area. Maximum day temperatures are around 32–35°C and minimum temperature 20°C. The second region includes all the area south of Cairo and has a mild almost rainless winter with maximum temperatures of 20–24°C and minimum of 5–9°C. The summer is hot with a maximum temperature of 36–42°C in daytime and 20–26°C at night. The climate is generally dry; only in August does humidity become relatively high.

Soils and Water Resources

The majority of the Egyptian soils presently cultivated are alluvial: level, deep, dark brown and heavy to medium in texture, constituting some 75% of the area under cultivation in the Nile Valley and Delta. Based on productivity, only 6% of the cultivated soils were classified as excellent and 45% as good, the remaining 49% were either medium or poor.

Egyptian agriculture is almost entirely confined to the Nile Valley and Delta and is wholly dependent on irrigation from the Nile River. Exceptions are a small area of irrigated land in the several depressions in the western desert where fossil ground water supplies water for irrigation, and a small area along the Mediterranean coast in the western desert and Sinai, with less than 200 mm rainfall, where barley and other minor crops are grown under rainfed conditions. Outside the Nile Valley and Delta and the Mediterranean coast, the bulk of Egypt (almost 97% of the total area) is an extremely arid desert.

Present Role of Legumes in the Farming System

Forage Legumes

Egyptian clover is a winter crop sown from late September to November following the harvesting of the summer crops of maize, rice and cotton, and is ploughed up in the spring and early summer in preparation for the summer crops. If clover precedes cotton, and by far the greater part of the cotton crop follows clover, which is more in the nature of a catch crop (Fig. 6), it is ploughed up early, usually in March, but when followed by other crops it goes for its full season, being ploughed up typically in May. Local types of clover have evolved corresponding with the two uses. One variety, Fahl, flowers early and is suitable for one cut. The other and most common type is Meskawi which allows several cuts to be taken because of its late flowering and branching characteristics.

When sown in early autumn, clover grows rapidly to give a first cut in 50 to 75 days and a second cut after an interval of 45 to 50 days. With increasing temperatures subsequent cuts become available in 30 to 40 days. Typically, full season clover will give four cuts and if sown early, five. Late sowings, in the cooler winter, take longer to reach the first cutting stage.

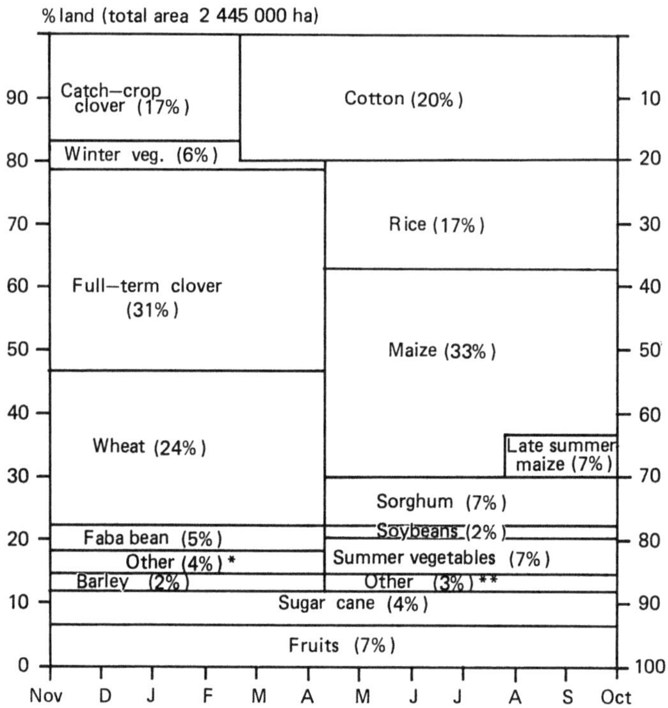

* Mainly flax, fenugreek, chickpea, lupin, onion, garlic, lentil, safflower and aromatic plants.

** Mainly groundnut, sesame, kenaf, summer forage and broom corn.

Fig. 6. Average land use (%) and pattern of crop rotations, 1980–82.

Together, full-term and catch-crop clover occupy 48% of the total winter cropped area (Fig. 6). This reflects the increased demand for forage as a result of the increase in the cattle and buffalo populations.

Clover is available for use from December to May and the bulk of the crop is cut by hand and fed fresh to all types of farm animals. Very little of the crop is made into hay. On large mechanized farms some ensilage is practised, but the high water content makes additives necessary. It is likely that there is so little hay or silage because on most of the farms on the old lands there is enough clover for the current needs of livestock, but no more. Also, near towns and large villages there is a ready market for the fresh crop, to feed horses and donkeys, and there is some trade between farmers.

The average expected yield of the green clover crop is widely estimated as being from 12 to 19 t/ha per cut depending on the stage of growth. A full-term crop of four cuts could yield 60 t/ha of green matter. The dry matter content would lie between 12 and 16%. The total annual production of the present 1.18 million ha of full-and catch-crop clover has been estimated at 50 million tonnes fresh matter or 7.5 million tonnes of dry matter. Recently-bred varieties produced 24–29 t/ha per cut. These varieties have been in distribution since 1986.

Seed production of clover is from farmers' own crops, the majority of farmers harvesting their own seed by reserving a suitable proportion of the last cut for this purpose. Foundation seed of the new varieties is produced in state farms of the Agricultural Research Centre. Maybe one to six thousand tonnes of commercial seed are exported annually.

Alfalfa is being grown extensively on big farms as well as on small owners' fields. It appears as early as possible in the reclamation cropping, generally after initial crops of Egyptian clover (berseem). If there are no unfavourable soil alkalinity and salinity problems, alfalfa will persist for many years. The crop is advocated as a means of improving soil fertility and as a basis for a program of animal production which is being introduced on the new lands. Under good management and full irrigation, alfalfa gives 8 to 10 cuts per annum, being least productive in the cool winter months. Average production is estimated at 10 t/ha of green forage per cut. Local varieties of alfalfa are more adapted to the sandy soil than the imported varieties.

Food Legumes

The four food legumes are winter crops sown for dry seed from mid October through November following the major summer crops, maize, rice, grain sorghum and cotton. Harvested in spring and early summer they are followed by those same crops in rotation (Fig. 6).

Faba beans are produced in Egypt for both green and dry consumption. To a considerable extent it is the same crop, but with a dual purpose. Early planting in September is common around large cities for green consumption. The crop may be left in the field for dry seed later on (Watson 1981). Green faba bean is grown on 15% of the area sown for production of dry seed and has thus become Egypt's second most important vegetable crop after onions. Average yield is estimated at 10.8 t/ha of green pods. The same varieties sown for dry seed could be used for green production, except in some areas near Alexandria where large-seeded varieties are grown for the latter purpose.

In South Egypt (Qena and Aswan governorates) farmers interplant faba bean (and sometimes lentil) between autumn-sown sugar-cane rows spaced at 90–100 cm. Crop competition is low since the rapid vegetative growth of legumes coincides with the slow growth of sugar cane. Interplanted faba bean yields are close to the solid planted ones; intercropped lentil yields are low, however, due to sensitivity to the frequent irrigation applied to sugar cane.

Farmers growing faba bean in rice-, cotton- and maize-cropping areas apply zero tillage to shorten turnaround time between crops, and save the cost of land preparation by planting faba bean on flat unplowed land following rice harvest, or on 'old' ridges remaining after cotton and maize harvest. Little or no seed yield differences were detected in on-farm trials between conventional and zero-tillage practices applied in faba bean production (Nassib and Basheer 1983).

Lentil was traditionally grown in the basin lands in south Egypt on the soil moisture residue after Nile flood water receded in autumn every year. Following construction of the Aswan high dam and the introduction of the canal irrigation system in these lands lentil acreage shrank due to competition from other cash crops and low yield (lentil is sensitive to over-watering and weeds). In the early eighties, lentil was largely introduced in rice- and cotton-cropping areas in the Nile Delta and the acreage gradually increased to reach 30% of the total lentil area in 1987. One or two irrigations are given to the crop in addition to rains. Two varieties were released to farmers. Promytrene herbicide was recommended for weed control.

Chickpeas are used in cooking and as roasted or candied snacks. The small white-coated seed varieties (kabuli) are more common than the large seeded ones. The crop area is in some districts of southern Egypt and in the east and west Nile Delta.

Peas are grown mainly in winter as vegetables on an estimated 72% of the total crop acreage and with an average yield of 14 t/ha of green pods. The crop is concentrated around cities and in some newly reclaimed areas in the Nile Delta.

In all four legumes the straw has economic value since it is fed to animals after it is cut fine by mechanical threshing.

Mechanization is widely used in land preparation before sowing and in chemical spraying for weed and insect control, as well as in threshing.

Research

Varieties of Egyptian clover selected by plant breeders from local material have been commercially introduced. Seed produced by individual farmers in small inter-mixed parcels of land would present difficulties in maintaining the identity of introduced varieties.

The production of new varieties at present includes selection for a high nitrogen fixation-type for local strains of *Rhizobium trifolii* for the Nile Delta and newly reclaimed sandy and calcareous soils.

A selection program among local ecotypes of alfalfa for yield, persistency and disease resistance is in progress. Two synthetic varieties were developed, Sewa synthetic and Nubaria synthetic. In faba beans, though much has been achieved in yield improvement and production stabilization at the national level, research is needed in breeding for more stable varieties less sensitive to environmental changes.

In the field of developing disease-resistant varieties a breakthrough was made in identifying sources of resistance to chocolate spot and rust diseases (Ibrahim and Nassib 1979; Khalil et al. 1984). Breeding material was raised and seed increase of some disease-resistant, high yielding lines is underway. Studies on pathogen virulence and genotype x pathogen interaction have to continue to control disease outbreaks.

Orobanche control packages were developed, on-farm tested, and recommended to farmers (Nassib et al. 1987). Higher levels of resistance to parasites than that of Giza 402 should be targeted by searching for material having different mechanisms for resistance and combining it with the one existing in Giza 402. New chemicals for orobanche control should be developed and tested.

Some variability in resistance to aphids was detected in lab and field screening. Standardization of techniques and evaluation methods is needed in both lab and field work to maximize association between the results obtained. This is necessary for an effective breeding program for resistance to aphids.

To increase national faba bean acreage, production costs should be reduced by multi-disciplinary investigations into the effects of zero and minimum tillage systems. Semi-mechanization for operations such as planting, harvesting and threshing should be developed to reduce turnaround time between faba bean and the following cotton crop. Together with developing early maturing varieties, this could help small farmers to increase cropping intensity and make better use of their limited resources.

The magnitude of faba bean green-pod production calls for strengthening research in breeding and development of agronomic packages to improve yield and quality.

The breeding program for lentil should be reoriented towards developing early maturing varieties better adapted to irrigation for planting prior to the cotton crop. Screening to identify sources of resistance against *Rhizoctonia*/*Fusarium* wilt is needed.

There has been some evidence that variability does exist in lentil genotypic response to local strains of *Rhizobium*. Since the crop is known to be a poor nitrogen fixer, efforts should be made to screen for genotypes efficient in nitrogen-fixation.

High costs of lentil harvesting discourage Egyptian farmers from planting the crop. Small harvesting machines need to be developed which might promote crop expansion.

In chickpeas, breeding strategy should aim at developing early maturing kabuli varieties resistant to the root rot and wilt complex, for traditional production areas. Since the crop could have some potential in light soils of the newly reclaimed areas under sprinkle irrigation, kabuli large-seeded varieties resistant to *Ascochyta* should be produced for export.

On-farm Research and Research-extension Linkages

Although on-farm research is well-known to agricultural research in Egypt, the new concept in verifying research results developed at stations under farmers' conditions and involving farmers themselves in the trials has been recently introduced through IDRC and ICARDA-IFAD Nile Valley projects on food legume improvement and USAID/EMCIP, Rice Research and Training, and NARP. The feed-back from such trials is very necessary in linking researchers with the actual crop production constraints. Efforts are underway to institutionalize the new concept of on-farm research in the research management structure of ARC.

The ICARDA/IFAD Nile Valley project, which has been implemented in Egypt and Sudan since 1979 and in Ethiopia since 1985, adopted on-farm research in verifying the effects of some test factors on faba bean production before calling farmers to take part in demonstrations.

In Egypt, test factors were varieties, tillage systems, plant populations, fertilizers, trace elements, irrigation, weed control, orobanche and aphid control. These were tried in two agroecologically-distinct zones for faba bean production. Socio-economic surveys of faba bean production and evaluation of the results of on-farm trials were conducted (Nassib and Basheer 1983).

After verification, packages including varieties, seed and fertilizer rates, weed, orobanche and aphid control were implemented by farmers in demonstration plots, then were monitored by the multidisciplinary research team and extension staff. Impact on production, information dissemination and farmers' attitudes towards recommendations were estimated.

The Subject Matter Specialists (SMS) of the Nile Valley Project were involved in a Training and Visit extension program with the district staff under the Menia-IFAD Agricultural Development project. Those in turn work as SMS in extending the recommended packages to village extension agents who through contact with farmers disseminate information to the rest of the village farmers.

Training of the whole national extension staff through regular meetings and dissemination of information through journals and bulletins is underway. More training and provision of facilities are needed to strengthen research-extension linkages and interactions for better service to farmers and improvement in agricultural production.

References

Ibrahim, A.A. and Nassib, A.M. 1979. Screening for disease resistance in broad beans (*Vicia faba*) in Egypt. FABIS 1:25.

Khalil, S.A., Nassib, A.M., Mohamed, H.A. and Habib, W.F. 1984. Identification of some sources of resistance for chocolate spot and rust in faba beans. Pages 80–94 *in* Systems of Cytogenetic Analysis in *Vicia faba* (Chapman, G.P. and Tarawali, S.A., eds), Nijhoff and Junk Publications, The Hague, The Netherlands.

Nassib, A.M. and Basheer, A. 1983. On-farm trials. Pages 38–47 *in* Faba Bean in the Nile Valley. Report on the First Phase of the ICARDA/IFAD Nile Valley Project (Saxena, M.C. and Stewart, R.A., eds). Martinus Nijhoff Publishers, Dordrecht, The Netherlands.

Nassib, A.M., Hussein, A.H.A., Saber, H.A. and Mousa, M.A. 1987. Orobanche control package in pilot demonstration plots. Pages 239–250 *in* Proceedings of the 12th International Congress for Statistics, Computer Science, Social and Demographic Research. Ain Shams University, 28 March-2 April 1987, Abbassia, Cairo, Egypt.

Watson, A.M. 1981. Faba bean production in Egypt. Study of the economic and infrastructural context. ICARDA, Aleppo, Syria.

Discussion

Durutan:
Which legume is the most profitable in Egypt (in terms of price)?

Nassib:
Probably berseem.

Ben Salah:
What is the yield level of Giza 402 under orobanche-infested conditions compared to orobanche-free conditions?

Nassib:
In one trial Giza 402 (tolerant) was evaluated in comparison with Giza 2 (susceptible) under severe orobanche infestation (100%). Giza 2 was completely wiped out whereas Giza 402 yielded 2 t/ha seed.

Webber:
How do you produce seed of Berseem clover – e.g. new cultivars – and how is it distributed to the farmers?

Nassib:
Seed production of the new varieties is still a problem. No special areas are available for production, even on state farms. Farmers produce their own seed, after cutting 2 or 3 times, usually in June.

Halila:
What kind of resistance to orobanche is there in Giza 402; mechanical or genetic?

Nassib:
Probably mechanical, due to more lignification in root cells which hinders the penetration of the haustoria. No genetic studies have been carried out so far.

Solh:
How does the root system biomass and spread of Giza 402 compare with that of other varieties in Egyptian irrigated conditions? In some locations under rainfed conditions Giza 402 has a comparatively small root system. Is this the reason for its higher tolerance to orobanche?

Nassib:
It does have a more compact root mass than other varieties, although this needs more study. This may be one of the reasons for its high tolerance of parasite infection.

Bahl:
To check the incidence of cotton leaf worm (*Spodoptera littoralis*) planting of alfalfa was prohibited by law in the delta area of Egypt. Was the planting completely checked, and was the incidence of leaf worm brought under control merely by not planting alfalfa?

Nassib:
Yes, planting of alfalfa was completely checked. In addition, the crop is probably not attractive to farmers as it is slow in growth in winter compared to berseem. Cotton leaf worm is still a serious problem: it has many host plants in addition to alfalfa and cotton.

Tawil:
What is the situation with soya bean in Egypt?

Nassib:
It began to be grown in the early seventies. Last season, 1986, the acreage was about 50 000 ha and the national average yield was 2.4 t/ha.

The Role of Legumes in the Farming Systems of Greece

C.I. PODIMATAS

Fodder Crops and Pastures Institute, 411–10 Larissa, Greece

Abstract. The total area cultivated with forage and food legumes in Greece is about 250 thousand hectares (10.45% of total arable area). The most important forage legumes are alfalfa and vetch, occupying 68 and 21% of the forage land, respectively. Food legumes occupy a little over 32 thousand hectares but are being cultivated less and less. Natural pastures, of which legumes constitute a considerable percentage, occupy just over five million hectares.

This paper discusses the role of each group of legumes, forage, food, and pasture, in the farming systems of Greece. Also discussed are the constraints and the future prospects for production of these legumes.

Introduction

The total area cultivated with forage crops in 1987 was about 352 thousand hectares (14.7% of the total forage area), of which 220 thousand hectares were sown with forage legumes. Alfalfa (*Medicago sativa*) is the most important, grown on 68% of the fodder legume land, while vetch, (*Vicia sativa*), occupies 21%. Other legumes, (*Vicia faba*, *Lathyrus cicera*, *Pisum sativum* etc), including annual clovers, were grown on the rest.

Alfalfa is grown for hay all over the country, with estimated yields of 12 tons/ha under irrigation, (70% of its total area), and 7 tons/ha without irrigation, while vetch, grown during the cool season in dry areas, has a mean hay yield of 8 tons/ha and a mean seed yield of 2.5 tons/ha.

The total area cultivated with food legumes in 1987 was only 1.36% of the total arable area. The allocation of different crops in this area is shown in Table 1.

All the above legumes are grown under rainfed conditions, except beans, which are irrigated.

In the past food legumes were very important for Greeks as their survival depended on them when food was short. They became traditional crops to complete or replace human needs for animal proteins because livestock supplies were unreliable. Important ecotypes evolved in many places. However, in the last 20 years, the gradual decrease in the area cultivated with food legumes, because of the different reasons given below, has obliged

A.E. Osman et al. (eds.), The Role of Legumes in the Farming Systems of the Mediterranean Areas, 63–69.

Table 1. Total area (ha) occupied by different food legume crops in 1987 and their average yield (kg/ha) 1977–87.

	Area	Average yield	Range in yield
Beans (*Phaseolus vulgaris*) grown alone	12 018	1470	600–4000
Beans with cover crops	6 859		
Chickpea (*Cicer arietinum*)	6 176	1120	600–3000
Faba bean (*Vicia faba*)[1]	4 317	1780	800–5500
Lentil (*Lens culinaris*)	1 735	1110	500–3000
Others	1 337	N.A.	N.A.
Total	32 442		

N.A. = not available.
[1] Grown for food and feed.

Source: National Statistical Service (1988).

us to import considerable quantities of beans, chickpea and lentil every year (National Statistical Service of Greece 1988).

With reference to grain forage legumes, we have been importing soya bean as a protein source for animal consumption. Last year, in order to replace the import of this expensive animal feed, it was cultivated in Greece, but the yields so far are very poor (mean yield 2500–3000 kg/ha). Soya bean has little potential for Greek agriculture: it is a spring crop, and thus requires irrigation. This brings it into direct competition with other cash crops which give a higher return, eg cotton, maize, sugar beet, and tomatoes (Roupakias 1983).

Natural pastures, including a large proportion of leguminous flora, occupy over 5 million hectares (40% of total area of Greece) and are the largest resource for the development of animal production (Fig. 1). Pure pasture,

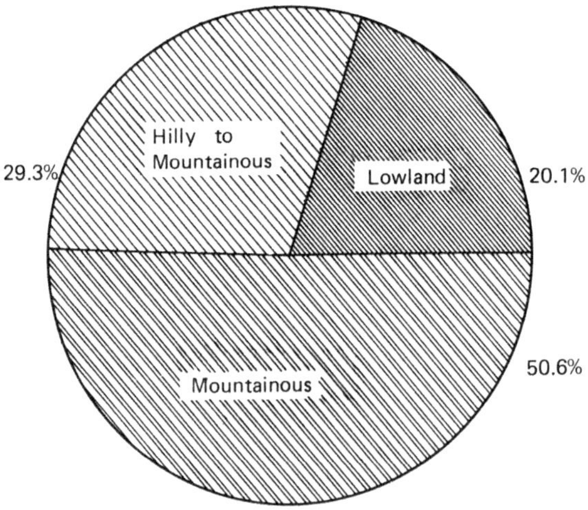

Fig. 1. Topography of natural pastures and relative area of each category (Total area 5 280 000 ha).

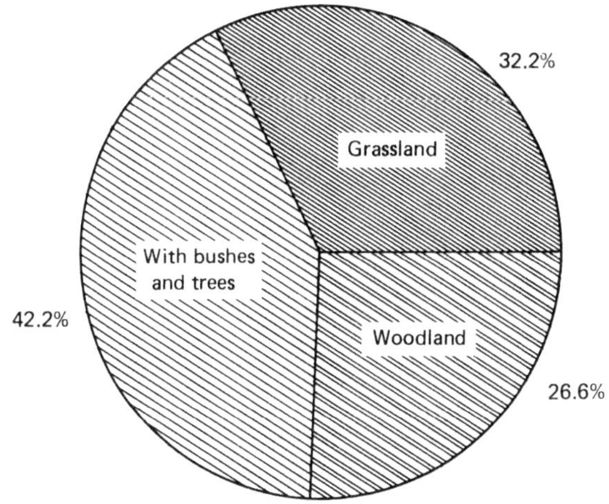

Fig. 2. Classification of natural pasture in Greece.

with or without scattered shrubs, is about 32% of the total area, the rest being mixed with trees and shrubs. Many years ago, much of these marginal areas was cultivated with forage and food legumes (Kontsiotou 1981; Vaitsis 1987). They are mostly in the hills or mountains, or in lowland sandy or rocky areas not considered suitable for cultivation nowadays (Fig. 2). Although there is not enough information about the classification and distribution of the flora of the natural pastures in the whole country, we can say that mountain pastures usually have a good mixture including mainly cool season perennial species. There is a large gap of green matter production in winter. *Trifolium repens*, *Trifolium hybridum* and *Onobrychis montana* are common. Pastures in the north-west and north-east zones also have good composition, including both warm and cool season perennials: *Trifolium pratense*, *Trifolium repens*, *Trifolium fragiferum*, *Lotus corniculatus*, *Onobrychis viciaefolia*, *Melilotus albus*, *Melilotus officinalis* and *Medicago falcata* are commonly found. As the altitude gets lower, the percentage of annual species increases: *Trifolium incarnatum*, *Trifolium subterranean*, *Medicago arabica* and *Medicago lupulina* are among the most important. In the south-east zone the percentage of annual species is high. Among these are *Trifolium hirtum*, *Trifolium subterranean* and *Medicago minima*. In this zone, there is a shortage of green matter production in summer.

Physical Environment

Climate

Four different zones can be distinguished (Fig. 3).

1) Mountainous zone: rather continental, characterized by low winter temperatures and dry, rather cool summers. There is high annual precipitation (600–1200 mm), including snow.

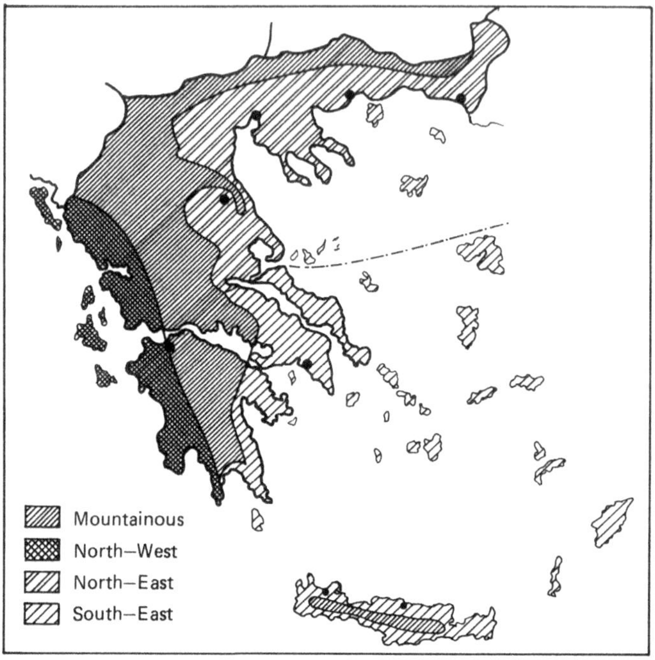

Fig. 3. Climatic zones of Greece.

2) North-west zone: mild winters and very dry summers, with 800–1000 mm annual rainfall.
3) North-east zone: similar to 1) in its climate, but precipitation is less (500–800 mm), including snow, and summers are hot.
4) South-east zone: typical Mediterranean climate. All rainfall occurs in the cool season, and summers are hot and extremely dry.

Present Role of Legumes in Farming Systems

The effect of several rotations with forage and food legumes on wheat yields under rainfed conditions has been studied in a long-term experiment started in 1937 at the Fodder Crops and Pastures Institute in Larissa.

The average grain yields obtained show that irrespective of the use of chemical fertilizers, wheat monoculture was inferior to wheat grown in different rotations with legumes. In particular, the green manure of a pea – wheat rotation resulted in high, stable wheat production right from the beginning, while the common vetch – wheat rotation gave a large initial yield increase which in time stabilized at a high level (Sotiriadis 1977). As a result of this trial and other observations, forage and food legumes were generally introduced into the rotations of wheat and other cereals in Greek agriculture during the 1950s and 1960s.

Subsequently, however, the farmers switched to crops more easily harvested by machine, mainly cereals. This move was actively supported by the Government, with the objective of attaining self-sufficiency in cereal production. In addition, with the movement of the agricultural population to the large cities, the livestock population fell (mainly horses, mules and asses), the marginal lands on hilly and mountainous areas were abandoned and the farmers switched to cereal monoculture as a more convenient crop (Kontsiotou 1981; Roupakias 1983).

As a result of all these facts, together with a change in Greek eating habits, the area of land cultivated with forage and food legumes has gradually declined over the last 20 years.

When we consider the perspectives, we believe that at present there is a need for a high protein animal feed to replace the importation of soya bean. There is also a need for self-sufficiency in the production of food legumes, such as beans, lentils and chickpeas. Furthermore, leguminous crops are needed in the rotation system of the rainfed area of Greece, which amounts to 56% of the agricultural land and which is mostly under continuous cereal cultivation at present. These could include faba beans, vetch and lentils, as winter-sown crops. Faba beans, in particular, are easy to harvest mechanically, and as a winter-sown legume, may be able to fulfil the needs for both high protein animal feed and the rotation. It is important here that faba beans should be successfully incorporated into the industrial process of making mixed animal feeds.

Finally, a very important factor for the use of the other food legumes in the rotation system of cereals is agricultural policy, which must favor the encouragement of legume cultivation.

Research

During the last fifteen years, a considerable research effort has been given in Greece to the breeding and management of legumes.

In plant breeding a number of new varieties have been developed which are registered in the national list. Also enough seeds have been produced for most of the leguminous crops. In future, our research should focus on the development of new varieties of legumes which are highly productive, more resistant to diseases and moisture stresses, and which are also suitable for mechanical harvesting.

We are satisfied that the most important cultural practices required by most of these crops have been adopted, and we are still working on effective new herbicides and use of fertilizers.

In addition, there has been research on the improvement of natural pastures. Six projects dealing with fertilization, weed control, management and reseeding of natural pastures were carried out in central and north-west Greece. The seeding involved single species and mixtures of annual and perennial grasses and legumes.

Analysis of the results gave us the following useful information:

- Fertilization, good management and weed control can significantly improve the composition of native flora and dry matter production and feed quality. The fertilization treatments recommended are 40–80 kg of N/ha and 50–100 kg of P_2O_5/ha.
- In most cases native flora yield at least as much as sown mixtures. Reseeding is therefore suggested only for pastures with infertile soil, where useful species have disappeared as a result of overgrazing or erosion.
- Most of the species studied were established easily and performed well. Native populations have always done better under hot, dry conditions than the introduced varieties, however (Vaitsis 1987).

References

Kontsiotou, H. 1981. Fodder legumes. Pages 67–78 *in* Proceedings of the 4th meeting of the FAO sub-network on Mediterranean pastures, Thessaloniki, Greece.

National Statistical Service of Greece. 1988. Monthly statistical bulletin, Vol. 33 No 1, 1988 Athens, Greece.

National Statistical Service of Greece. 1988. Pre-census data of the agriculture-livestock census for 1987. Athens, Greece.

National Statistical Service of Greece. Statistical Yearbook of Greece, 1978–1988. Athens, Greece.

Roupakias, D.G. 1983. Faba beans in Greece: past and future. FABIS Newletter No. 7: 6–8. ICARDA, Aleppo, Syria.

Sotiriadis, S.E. 1977 The trend of differentiation of wheat yields in relation to the rotation used. Agricultural Research I: 125–136, Ministry of Agriculture, Athens, Greece.

Vaitsis, T.A. 1987. Grasslands and fodder crops in Greece. Pages 185–191 *in* Agriculture: Pasture Improvement. EEC, Madrid, Spain.

Discussion

Solh:
With the adoption of cereal monoculture did farmers increase inputs or improve management?

Podimatas:
The farmers increased inputs significantly, mainly N-fertilization, but some years after the adoption of this monoculture we already have problems with the decrease in pH of the soil because of the continuous use of NH_4SO_4.

Solh:
What has been the trend in the productivity of cereals after changing to monoculture?

Podimatas:
It is decreasing. However, it must be said that there would be problems changing back again to cultivating legumes, because of difficulties in mechanization, productivity etc.

Abd El Moneim:
What is the rainfall in vetch-growing areas?
Do you use the same cultural practices for both hay and grain production? Do farmers produce their own seeds?

Podimatas:
600–700 mm rainfall
They are the same, except for the sowing rate: for hay this is 140 kg/ha and for grain 120 kg/ha.
Greek farmers do not produce their own seeds. The state is responsible for seed production of all varieties registered on the national list, under the control of the Ministry of Agriculture. Five years ago the EEC asked us to develop a common policy on official seed production. The price of certified seed, for all varieties on the national list, is fixed by the government each year.

The Role of Legumes in the Farming Systems of Iraq

A.H.K. AL-ANNEY

Forage and Legume Research Section, Department of Field Crops General Board for Agricultural Research and Water Resources Ministry of Agriculture and Irrigation, Baghdad, Iraq

Abstract. Legumes are not some of Iraq's most important crops, although because of increasing demand for more animal and plant protein, the area grown with legumes, especially forage legumes, is expanding. They are mostly grown in winter, for their green matter, or dry seeds, or both. More money and effort are being invested in agriculture to reduce dependence on food imports. Legume research includes projects on selection of cultivars, response to fertilizers, and other management aspects. There is cooperation in research between the Field Crops Department and ICARDA, as well as other regional and international organizations.

Introduction

A very large proportion of Iraq is desert or mountains, so that out of its 43 million hectares only eight are able to support agriculture, and of these only half are cultivated in any one year. In the rainfed zone winter cereals, wheat and barley, are the main crops, with some legumes, and in the irrigated zone wheat is grown in winter, and rice, cotton and vegetables in the summer. Food and forage legumes are grown in all parts of the country, but are concentrated in the central and northern regions. They make a relatively small contribution to agricultural production at present, reflecting the general lack of integration between the arable and livestock sides of traditional agriculture. The most common food legumes are, in order of importance, faba bean, chickpea, lentil, and pea, with alfalfa, berseem, vetch and medics the most often grown forage crops. Animals – four-fifths of the 20 million livestock are sheep and goats (FAO 1984) – tend to be fed on fallow and permanent pasture rather than on especially grown forage. Fig. 1 and Table 1 show the area and production of some of these legumes, 1976–86.

Constraints to agriculture include salinity and waterlogging in the irrigated areas, unreliable rainfall, extremes of temperature, insect pests (locusts), and erosion of the soil by water (15% of the land is at risk) and wind (75% at risk). Everywhere there is a shortage of labour, especially skilled management.

A.E. Osman et al. (eds.), The Role of Legumes in the Farming Systems of the Mediterranean Areas, 71–75.

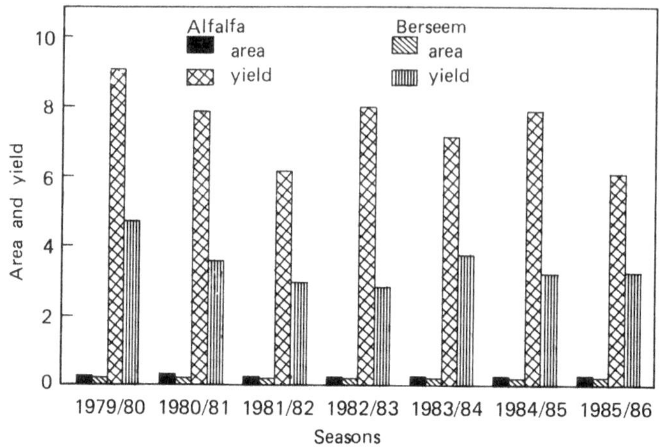

Fig. 1. Area (1000 ha) and total yield (1000 t, green matter) of alfalfa and berseem 1979–86.

Table 1. Area (1000 ha) and total production (1000 t) of dried seed of faba bean, chickpea and lentil 1976–86.

Seasons	1976 /77	1977 /78	1978 /79	1979 /80	1980 /81	1981 /82	1982 /83	1983 /84	1984 /85	1985 /86
Faba Bean										
Area	15	15	11	16	16	16	7	5	6	6
Prod.	16	15	11	17	17	17	7	8	8	9
Chickpea										
Area	17	14	18	14	14	15	15	16	14	16
Prod.	11	7	12	17	9	9	9	9	8	13
Lentil										
Area	6	10	7	10	10	10	10	5	5	6
Prod.	6	8	7	9	9	9	4	3	5	6

Physical Environment

Four agro-climatic zones can be distinguished (FAO 1959):

1. The Mesopotamia valley, between the Tigris and Euphrates rivers (150,000 km^2). The Lower Plain is never higher than 30 m above sea level, has extreme temperatures, ranging from an average 10°C in January to 50°C in summer, and little rainfall, from 5 to 200 mm, rising to 300 mm at the eastern extreme, concentrated in winter. Soil consists of clay and calcareous loam, with sandy, swampy, saline areas in between. The Upper Plain rises gradually in the north to 300 m above sea level,

and consists of dry steppe and desert. Rainfall is between 100 and 300 mm, and temperatures range from 7°C to 50°C. The growing season is limited to March and April. Calcareous grey soils alternate with gypseous areas and gravel deserts.

2. The Kurdestan mountains (25 000 km^2). Winters are severe with frequent frost and snow, while summer temperatures can rise to 40°C. Average precipitation is 600–1400 mm. The soils are mainly reddish chestnut, brown forest, red Mediterranean and associated lithosols. Oak forests and alpine pastures are characteristic of this region.
3. The foothills, a belt of undulating hills with altitudes of between 200 and 450 m (40 000 km^2). Rainfall varies from 300–600 mm. Winter temperatures can drop to −10°C, with a January mean of 7–8°C, but they rise sharply in March. Soils are mostly brown and reddish-brown gypsiferous, with some loess and alluvium. Rainfed cereals are grown here, but over the years overgrazing and poor farming practices, such as prolonged fallowing, have diminished productivity. Soil erosion is a serious problem.
4. Desert. Very low rainfall of 50–200 mm can only support subsistence agriculture round scattered wells. It covers an area of 250 000 km^2, more than the other three areas put together.

Present Role of Legumes in Farming Systems

Human and Animal Nutrition

Food and forage legumes help to supply the increasing demand for more plant and animal protein. For human consumption faba beans and cowpeas are used either as green or dry seeds, but only dry chickpea or vetch seeds are used. They are used in many traditional dishes, irrespective of incomes, especially in winter and during the Ramadhan fast. Animals eat the green matter or the poorer quality dry seeds. Forage legumes are sometimes grown as a cereal-legume mixture, especially in areas where there are a number of large animal-raising projects and where several cuts can be taken each season. The seeds are rarely used in a mixed-feed formula, with the exception of vetch.

Crop Sequence

The traditional 'Nirin' system follows a two-year rotation, winter wheat/barley-fallow. Summer cropping is minimal, using only 10% of the land, growing sesame, cotton, vegetables and green cereals. Fallow is grazed, and may be irrigated a few times to promote weed growth. Cereal-legume rotations are practised by only a small number of farmers, although a few research projects on forage and legume rotations are in progress.

Mechanization

Since most of the farms that cultivate legumes are small, there is little mechanization. Small tractors may be used to plough the land before sowing. Research is in progress on the selection of erect legume crops suitable for mechanised harvesting.

Socio-economic Considerations

For a while government policies shifted towards development of urban, non-agricultural sectors, so that by 1985 agricultural products formed only 10% of the GDP, and half the rural population had moved to the towns, whereas in 1960 agriculture was the single most important component of the Gross Domestic Product (GDP), with 74% of the population living in rural areas (Arab Agriculture Yearbook 1988).

However, policies are having to change: 60% of Iraq's food requirements were imported in 1987. Also, the rise in per capita income, and the improved education in nutrition, have increased the demand for protein foods, i.e. legumes, and livestock products, including feed and forage to feed that livestock. There are efforts being made to make better use of the existing cultivated areas through providing farmers with fertilizers and extension. Since 1960 purchase and marketing have been controlled by the government. Price incentives are working well, and marketing by the private sector is being liberalized to encourage farmers to increase output. After land reforms of 1985 most farms are state- or state agency-run. Only in the northern mountain areas is there large-scale private ownership of land. The government is also investing in large-scale irrigation schemes and land reclamation.

Research and Extension

In general, because of Iraq's dependence on imports for so much of its food, more attention is being given to improving agricultural production. In September 1987 the government commissioned a detailed study of agricultural and agro-industrial projects feasible in Iraq, by the Khartoum-based Arab Authority for Agricultural Investment and Development. Range management was included in the National Development Plan 1976–80, with projects to develop sustainable agriculture in the western desert by soil conservation, and the establishment of range and feed storage stations and fenced areas to demonstrate the advantages of range management. There were also plans to combat desertification.

Large scale dairy farms are to be developed, with their own animal-feed factories and forage farms.

In this resurgence of agricultural effort legumes should play an important part, in both human and animal nutrition, and in checking soil erosion and improving the soil generally. Research into all these aspects is necessary, and is already being carried out, mainly in the Field Crops Department, but also in the Agricultural Colleges at the Universities of Baghdad and Mosul, as thesis projects. Research has focused mainly on selection of suitable cultivars of the most important food and feed legumes, but there are also projects on date of sowing, spacing, legume-cereal mixtures, different types and levels of fertilizers, and seed inoculation. ICARDA sends seeds of a number of cultivars for testing against local cultivars, including medics. The eight agricultural research stations play an important role in screening cultivars.

References

FAO Production Yearbook 1983. 1984. UNFAO, Rome, Italy.
Arab Agriculture Yearbook 1988. Arab World Agribusiness, Falcon Publishing WLL, Bahrein.
FAO Mediterranean Development Project. Iraq country report. 1959. UNFAO, Rome, Italy.

The Role of Legumes in the Farming Systems of Jordan

N.I. HADDAD and B.A. SNOBAR

Faculty of Agriculture, University of Jordan, Amman, Jordan

Abstract. Lentil, chickpea and bitter vetch are the major legume crops grown in the rainfed areas of Jordan. They are a major component in crop rotations. Special attention was given recently to forage legumes in the five year economic and social development plan. Now more research and demonstrations are being conducted. As a result, some farmers have replaced fallow with medics or forage mixtures (barley and common vetch) for animal grazing. The major problem facing production of legumes at the farmer level is the lack of mechanization, especially the mechanization of harvesting.

Introduction

Jordan is an agricultural country with approximately 0.7 million hectares of arable land. Rainfed agriculture covers 490 thousand hectares while only 3500 hectares are under irrigation.

The importance of the agricultural sector stems from its being the main source of income for about 20% of the labor force. Although there has been noticeable growth in the output of fruits, vegetables and livestock products, the agricultural sector has not been able to meet the increasing demands, especially for meat and dairy products, resulting from population growth and greater spending power.

This increased demand for food in Jordan has led to increased dependence on imports of food and livestock products. Such imports have increased from an annual average value of JD 41 million, 1973–1975, to JD 94.4 million, 1976–1980, and to JD 179.9 million, 1981–1985, representing 16.5% of total imports during that last period (Ministry of Planning 1986).

Legume Crops

Lentil and chickpea are the major food legume crops planted in the rainfed areas of Jordan. Faba bean is usually grown under irrigation and mainly in the Jordan Valley, mostly for fresh pod utilization. The major forage legume grown in the rainfed areas is bitter vetch (*Vicia ervilia* L.). This crop is produced mainly for its seeds and straw (tibin), which are used for animal

A.E. Osman et al. (eds.), The Role of Legumes in the Farming Systems of the Mediterranean Areas, 77–83.

feeding. The area harvested, production and yields of lentil, chickpea and vetches are presented in Table 1.

There has been a general decrease in area planted to lentil, chickpea and vetches over the past 14 years, as well as in total production of these crops. The overall decline in area appears to be the result of the general increase in production costs, especially labor for harvesting. The poor yields are considered to be due to the traditional methods of production and the low-yielding landraces usually used by farmers. However, a contributing factor to the poor yields of these crops is the general trend by farmers to use the more productive land for wheat and the less productive stony land for the legumes. Yields per hectare fluctuate from year to year, correlating with seasonal variations in rainfall. The total production of lentil, chickpea and vetches does not meet the demand and as a result there are substantial imports.

The area planted to other forage legumes such as alfalfa, clovers, medics, and common vetches (Bekia) is still small but is increasing each year. Alfalfa and clover are mainly grown under irrigation while medics and vetch are rainfed.

Grazing in Jordan usually takes place from December to June. The sheep graze grass during winter and spring, and cereal stubble and crop residues during the summer season. Sheep generally walk between 3 to 10 km daily for grazing (Sawafta 1985).

Physical Environment

Jordan has a Mediterranean climate, with warm dry summers and mild winters. Rain falls during December and May, December to March being

Table 1. Area (1000 ha), production (t) and yield (t/ha) for lentil, chickpea and bitter vetch, 1973–86.

	Lentil			Chickpea			Bitter vetch		
Year	Area	Prod.	Yield	Area	Prod.	Yield	Area	Prod.	Yield
1973	24	4827	0.20	7	1891	0.26	8	2296	0.29
1974	22	31444	1.46	12	9101	0.73	8	6883	0.88
1975	15	5238	0.35	4	949	0.27	4	1959	0.46
1976	23	9380	0.41	2	358	0.22	5	1836	0.39
1977	13	5971	0.44	1	562	0.41	4	1906	0.43
1978	14	8345	0.57	1	352	0.28	4	2957	0.67
1979	7	811	0.11	3	417	0.16	4	1086	0.25
1980	9	6295	0.73	3	1654	0.58	2	1778	0.76
1981	11	7872	0.75	2	1484	0.75	3	2493	0.76
1982	11	8084	0.75	2	1497	0.74	3	2573	0.76
1983	6	5173	0.88	2	678	0.44	2	2076	0.89
1984	5	2480	0.51	1	612	0.46	3	2127	0.62
1985	6	4063	0.70	3	1589	0.55	4	2309	0.65
1986	3	1750	0.61	1	593	0.51	1	1270	0.92

the wettest period. Agroecologically, the country can be divided into five major zones:

1. Desert area (91%); rainfall <200 mm.
2. Marginal areas (6%); rainfall 200–350 mm.
3. Semi-arid areas (about 2%); rainfall 350–500 mm.
4. Semi-humid areas (1%); rainfall 500–800 mm.
5. The Jordan Valley; rainfall 300 mm in the north decreasing to 100 mm in the south. The area has a hot climate and most of the land, between 200 and 350 m below sea level, is under irrigation.

Field crops are mainly planted in the marginal and semi-humid areas.

Present Role of Legumes in Farming Systems

Lentil, chickpea and faba bean are a major protein source in the diet of the Jordanian people, and are a main ingredient in several dishes. Although the amount consumed per capita remains the same the total amount consumed is increasing yearly because of the increasing population. The demand for forage legumes is also increasing each year because of the greater numbers of livestock, mainly sheep and goats. In rainfed areas, lentil, chickpea and vetch are the most common legume crops planted in the rotation with wheat and barley. Lentil and vetch are planted in the autumn and are considered winter crops; chickpea is planted in the spring and is considered a summer crop.

Crop rotations of either two- or three-year duration are practised. The majority of farmers include fallowing in the rotation, especially in the low rainfall zones. Such a rotation will generally consist of cereal (barley or wheat)-fallow; a legume crop is rarely included in the low rainfall zone. In the relatively high rainfall zones (>300 mm), however, a two-year rotation of either wheat-lentil or wheat-vetches is practised. A considerable area under a wheat-fallow rotation system is also found in the high rainfall zones. A three-year rotation is now practised by several farmers in high rainfall zones. The crops included in this rotation generally consist of wheat followed by lentil followed by a summer crop, either tobacco, summer vegetables or chickpea.

Recently, some farmers have replaced fallow with medics or a forage mixture (barley and common vetch) for animal grazing. This system has become popular in the central part of Jordan, where livestock production is an important component in the farming system.

A survey of 25 farmers in Madaba district (south of Amman) revealed that 28% of the farmers used a cereal-fallow rotation, 32% a cereal-forage legumes rotation, 20% a cereals-summer crop rotation, 12% a cereal-food legume rotation, and only 8% followed a 3-year rotation of cereals-forage legumes-summer crops (Sawafta 1985).

At present, the most serious problem facing the production of legumes at the farmer level is the lack of mechanization; especially the mechanization of harvesting. Harvesting is still a hand labour operation that accounts for more than 50% of total production costs (Haddad and Arabiat 1985). If legume mechanization is successful, a significant increase in the area planted to legume crops can be expected.

Research and Extension in Legumes

Active research in food legumes started at the University of Jordan (UJ), Faculty of Agriculture in 1980. The research activity was strengthened by financial support from the UJ and the International Development Research Center (IDRC). The research covered several areas and included the identification of proper cultural practices for growing lentil and chickpea. Also essential for lentil and chickpea production is adequate control of weeds, insects and diseases. Breeding high yielding cultivars with wide adaptation, the ability to remain erect at maturity, and acceptable seed quality is also essential. Crop mechanization, especially harvest mechanization, was given special emphasis in this program (Haddad 1983; Haddad 1987). The above research activities have resulted in a set of recommendations that were published in two extension bulletins addressed to extension agents and farmers.

The project was also designed to provide training for Jordanians in the area of legume research and production, and as a result, short and long term training programs were provided by ICARDA and ICRISAT to project personnel. Moreover, the project provided scholarships to nine Jordanians who worked on food legumes. Some of those graduates are currently involved in research and extension work on legumes in the Ministry of Agriculture or other organizations in the country.

Research work in forage legumes was given special attention in the previous five year economic and social development plan (1980–85), and has continued in the present development plan (Ababneh 1983). The plans emphasize the introduction of forage legumes into the farming systems of Jordan. They include the introduction of forage mixtures (cereals and legumes) to the high rainfall areas, and the introduction of vetches and medics to the marginal areas. The work is conducted in cooperation with the Ministry of Agriculture, Faculty of Agriculture, UJ, and the Jordan Cooperative Organization. The activity is partially sponsored by UNDP and the Australian Government. Research activities continued for about six years and have covered different aspects of production practices with the integration of livestock into the systems. Grazing trials were conducted and farmer participation was taken into account from the start (Bull 1983).

A project recently initiated between the Ministry of Agriculture and the Arab Center for the Studies of Arid Zones and Dry Areas (ACSAD), has taken an integrated approach to crops and livestock systems. In this project cropping systems that contain cereals (wheat or barley), food legumes (lentil), forage legumes (common vetch or medics) and fallow, are being evaluated in marginal areas of the Ramtha Research Station and in semi-arid areas.

The three-year study is designed to investigate the effect of the different systems on soil moisture, soil nutrients and on the performance of the succeeding cereal crops. It will also investigate the potential of forage legumes for animal fattening. Forage quality of both medics and common vetch will also be evaluated.

Because of the importance of legumes in the cropping system of the rainfed areas of Jordan, several studies were conducted by graduate students working toward their M.Sc. degrees. Their major area of research included the evaluation of legume species and optimum cultural practices, legume-cereal mixtures, legumes in the crop rotation, and pasture forage legumes for lamb fattening. Special emphasis was given to plant quality (Ababneh 1983; Bull 1988; El Masri 1988).

It is obvious from the above review that there is agreement between all agricultural organizations in the country on the importance of food and forage legumes in the cropping system and crop rotations of the rainfed and irrigated areas of Jordan. This was clearly emphasized in the on-going five year economic and social development plan (1986–1990). Moreover, the newly developed cropping system for the Jordan Valley encourages farmers to grow forage legumes instead of vegetables. Along the same lines, the government's policy on investing in the newly developed projects in the South-east of Jordan using underground water, is emphasizing the production of forage legumes and livestock as a major part of the cropping system in those areas.

It is hoped that the available technology, recently developed in Jordan regarding the production of food and forage legumes, will be widely demonstrated to farmers to encourage its adoption, which is the ultimate goal. It is also hoped that new research areas will be developed and that the present research and demonstration activities will continue.

References

Ababneh, H.M. 1983. Studies on barley-forage legume mixtures under rainfed conditions in Jordan. M.Sc. thesis. Faculty of Agriculture, University of Jordan, Amman, Jordan.

Bull, B.C. 1983. Jordan-Australian dryland farming project. Annual report No. 3, 1982/83. Amman, Jordan

El-Masri, T. 1988. Effect of forage legumes on the production of succeeding wheat crop. M.Sc. thesis. Faculty of Agriculture, University of Jordan, Amman, Jordan (In preparation).

Haddad, N.I. 1983. Food Legume Improvement Project in collaboration with IDRC. Faculty of Agriculture, University of Jordan, Amman, Jordan.

Haddad, N.I. 1987. Food Legume Improvement Project in collaboration with IDRC. Faculty of Agriculture, University of Jordan, Amman, Jordan.

Haddad, N.I., and Arabiat, S. 1985. Problems and methods of lentil production in Jordan. Dirasat. Vol 12(6): 51–87. (In Arabic).
Ministry of Planning, 1986. Five year economic and social development plan, 1986–1990. Hashimite Kingdom of Jordan. National Press, Amman, Jordan.
Sawafta, I.Y. 1985. Production of forage crops, and socio-economic factors affecting sheep production under rainfed conditions in Madaba district. M.Sc. thesis. Faculty of Agriculture, University of Jordan, Amman, Jordan.

Discussion

Halila:
What is the latest on lentil harvest mechanization? Will attempts to develop machinery continue, or is there any prospect of identifying some genetic material well-suited to mechanical harvesting?

Haddad:
We have not succeeded in mechanizing lentil harvesting at the farmer level. The problem is not just one of variety or machine, but of seedbed preparation, seeding and general crop management. At the research level we had some success. In our breeding program at the University of Jordan we found that tall upright lentil genotypes are generally low yielders, due to the small number of branches they produce, and are prone to lodging, because they bear pods at the top of the plant. We are looking now for genotypes that have many branches that help to maintain a non-lodging stand and will be suitable for harvesting mechanically. These genotypes have high yield potential generally. Several are under evaluation in our own program.

Durutan:
Most of the time it is soil tillage which is the main factor responsible for low yields in a legume-wheat rotation. Even in areas where ecology permits the elimination of fallow, incorrect soil tillage reduces the wheat yield. Research should be carried out considering this problem too.

Haddad:
I agree. In Jordan a little work is being conducted on the effect of soil tillage within a rotation but still more work is urgently needed.

Papastylianou:
Vetch is known to be bitter at early stages of growth and it cannot be grazed, while it is good for hay-making. On the other hand, medics are suitable for grazing but not for hay-making. How do you compare them in your studies?

Haddad:
Several studies are in progress comparing vetch and medic for direct grazing. Others are investigating the potential of vetch for hay. Results of these studies are not yet finalised.

Gintzburger:
What about the validity of comparing vetch and medic as they appear to fit into the different farming systems?

Haddad:
They are testing both medics and vetches for direct grazing, so both of them are aimed to fit into the same type of system. There will be an economic evaluation of the net return from each situation. This work is not just conducted in low rainfall areas but also under relatively high rainfall. I believe vetch, which needs to be seeded every year, might give some flexibility to the farmer if for one reason or another he decides not to grow forage legumes for 1 or 2 years. This also needs further evaluation.

Robson:
What experiments have been conducted to look at the effect of fallowing after legumes on the yield of cereals?

Haddad:
None so far, but this is an important issue which should be studied.

The Role of Legumes in the Farming Systems of Morocco

M. BOUNEJMATE

INRA, B.P. 415, Rabat, Morocco

Abstract. Legumes are grown on an area of about 700 000 ha (nearly 10% of the total arable land), of which 70% is devoted to food legumes (faba beans, lentils, chickpeas, peas) and the rest to forage legumes (alfalfa, berseem, vetch). In the past 10 years, the area has increased significantly but production has remained the same for food legumes. The consumption of food legumes and the amounts exported decreased tremendously.

Legumes are mainly grown in areas with more than 400 mm annual rainfall or under irrigation. Poor farming practices (low quality seeds, incorrect fertilization, absence of pest control, etc) causing low yield, and low prices limit the cultivation of legumes. Recently, there has been increasing interest in the introduction of legume pastures (mainly in areas of 250–500 mm rainfall) and winter chickpeas. Research with more emphasis on on-farm research and development activities has also been intensified. It is likely therefore that the area sown to legumes will continue to expand with significant increase in yields.

Introduction

Legumes have been cultivated in Morocco for a long time and are the second most important crop after cereals, occupying about 10% of the total arable land. The trends in area and production for the past 10 years are presented in Table 1.

Food legumes are mainly cultivated under rainfed conditions. Faba beans (with approximately half of the overall acreage), peas, lentils and chickpeas are the major food legume crops. The area is subject to annual fluctuations. Between 1979 and 1987, the area has increased (mainly due to the increase in lentil area) whereas the production has stayed almost steady. The average yield over the past 10 years is 750, 460, 620 and 600 kg/ha, for faba beans, lentils, chickpeas and peas, respectively.

Despite an important increase since 1979, forage legumes occupy only a small area (nearly 200 000 ha in 1987). Alfalfa and berseem under irrigation and vetch (in mixture with oats) under rainfed conditions are the main forage legumes cultivated with an average yield of 44, 45 and 18 tons/ha fresh matter, respectively.

Until 1976, Morocco was the second country in the region in terms of

A.E. Osman et al. (eds.), The Role of Legumes in the Farming Systems of the Mediterranean Areas, 85–92.

Table 1. Trends in acreage (1000 ha) and production of food legumes (100 t grain) and forage legumes (1000 t fresh matter) 1978–1987.

Season		Food legumes						Forage legumes						
		Faba bean	Pea	Lentil	Chickpea	Others	Subtotal I	Alfalfa	Berseem	Vetch	Pea	Medics	Subtotal II	Total
1978–79	A*	208	63	29	62	91	453	44	21	16	15	0	96	549
	P*	1475	641	135	642	434	3327	2064	943	398	339	0	3743	
1979–80	A	156	47	37	66	79	384	51	21	14	9	0	95	479
	P	1044	273	160	446	387	2310	2701	1032	257	113	0	4104	
1980–81	A	130	36	34	32	71	304	56	26	20	9	0	111	415
	P	388	9	47	61	162	667	2340	992	143	37	0	3512	
1981–82	A	111	34	39	61	50	295	71	33	63	16	0	183	850
	P	987	218	253	509	250	2216	3112	1615	1400	335	0	6462	
1982–83	A	171	48	79	65	50	411	71	34	75	23	0	203	614
	P	1422	222	320	559	329	2852	2517	1422	1426	312	0	5678	
1983–84	A	190	57	68	60	68	444	62	33	83	13	0	191	635
	P	1223	276	245	287	425	2456	2623	1740	1271	177	0	5811	
1984–85	A	212	57	88	77	78	512	62	35	77	14	0	188	700
	P	1995	482	462	454	543	3885	2741	1719	834	168	0	5461	
1985–86	A	196	50	86	82	88	501	61	40	70	10	20	200	701
	P	2146	419	696	705	715	4680	2877	2160	1788	212	0		
1986–87	A	211	45	90	77	107	530	62	49	62	6	15	195	724
	P	1273	222	356	613	673	3137	2843	2203	1088	77			

* A = acreage; P = production.

Source: Production des legumineuses alimentaires au Maroc. Journees d'etudes, Settat. Amine M. and El Adlouni F. (1987).

food legume export. About 0.2 million tons of faba beans, chickpeas, lentils and peas were exported each year. After 1976, due to the decrease in production, almost all were consumed locally, with only small amounts being exported (Fig. 1).

Physical Environment

Morocco is divided into 15 ecological zones in relation to soil and climate (Fig. 2). Calcareous soils of fine to medium texture predominate. They are of low nitrogen and phosphorus content but potassium content is high. Rainfall in the agricultural zones varies from 200 to 1000 mm. The minimum mean of the coldest month varies from 0°C to 8.6°C.

Food legumes are mainly grown in areas with more than 400 mm annual rainfall, with deep clay soils. More than 85% of the area is located in zone 2 (Taounate, Taza, Fes, Meknes), with the remaining area in zone 8 (Settat, Safi).

Rainfed forage legumes are also mainly grown in zone 2 (vetch) and zone 8 (peas). Irrigated forage legumes are more widely spread. Alfalfa is grown in zones 4, 6, 8, 9 and 10 while berseem is cultivated in zones 3, 4, 8 and 10.

In some areas, environmental constraints limit the growth of some legumes. For example, due to waterlogging, alfalfa is not cultivated in the Gharb (zone 3). Nodulation failures occur with berseem in zones 6′ and 3.

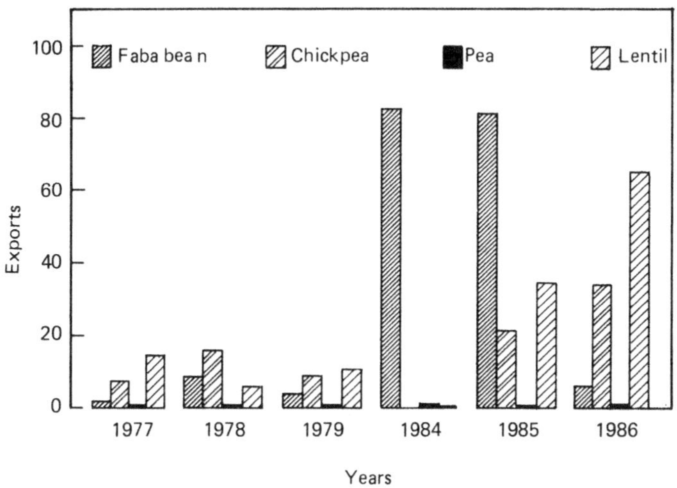

Fig. 1. Trends in exports of food legumes (1000 t).

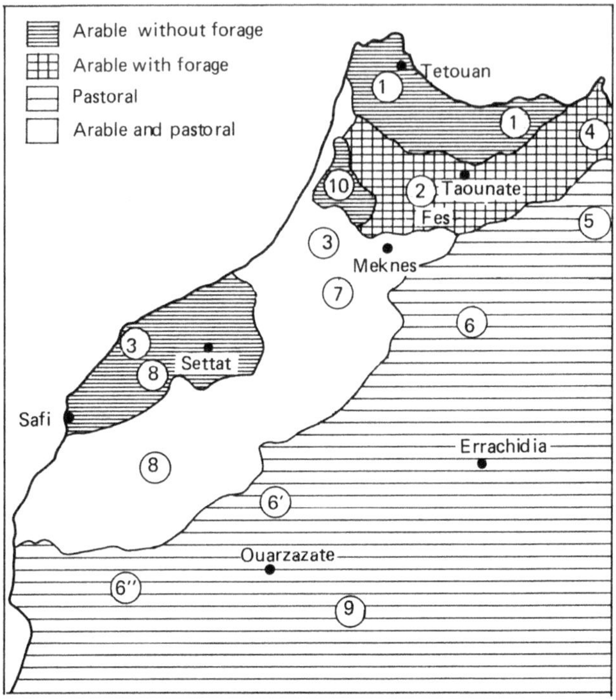

Source: MARA/FAO, TCP/MOR/4402, 1986.

Fig. 2. Ecological zones of Morocco.

Present Role of Legumes in Farming Systems

Human and Animal Nutrition

The consumption of food legumes is decreasing, falling from 9 kg/head/year in 1960 to 5 kg/head/year in 1971, mainly because production has remained the same and the population has increased. The amounts consumed vary between regions. Some regions are well known for their high consumption, e.g. the North. Faba beans represent 50% of the amounts consumed.

The contribution of legumes to animal feeding is increasing, owing to the increase in area: about 10% of animal feed comes from legumes:

Grains (faba beans, *Vicia ervilia*)	= 70 000 tons = 0.8% of total livestock needs
Straw and stubble	= 400 000 tons = 1% of total livestock needs
Forage (green or hay)	= 6 million tons = 8% of total livestock needs

Crop Sequence

The area devoted to cereals represents 4 to 5 million hectares/year. Thus the cereal-based system predominates. In dry areas, barley-barley and barley-fallow are the main rotations used. Peas and lentils are grown to a much lesser extent in rotation with barley or wheat. In wet areas, wheat replaces barley in the rotation. In these areas legumes are used in a two year rotation with wheat: wheat-faba beans, wheat-peas, wheat-vetch. In some cases, legumes are used as a green manure before wheat (berseem and lupin).

In irrigated areas, the rotations used are diversified. The crops following legumes (alfalfa, berseem and beans) are numerous: cereals, cotton, sugar beet, vegetables etc.

Crop Livestock Interaction

The integration of crops and livestock is an ancient and common practice, especially in the arid and semi-arid areas. Livestock are kept as a way of spreading the risk due to rainfall variability, but also to maximize the utilization of resources on the farm (fallow, crop residues, grains) and on the common rangeland. At the same time livestock play the role of a bank.

In irrigated areas, milk production from cows is almost always associated with crops. Money from milk sales is always available to finance small routine expenses.

Research

Constraints Limiting Productivity

Due to poor farming practices, namely cultivation and management, yields are low and represent less than 50% of the expected yields. Low quality seeds, rough soil preparation, incorrect fertilization, absence of pest control, insufficient mechanization and low prices are the main factors which limit the cultivation of legumes (Table 2).

Legume Research Programs

Table 3 presents the on-going work on legumes. Three national organizations are involved in cooperation with some international centers. The experiments deal with various topics. They attempt the development of superior cultivars and management practices, and the adaptation to local farming systems of winter chickpeas recently released by ICARDA. In cooperation with ICARDA, an ecogeographic survey of native forage legumes started this year.

Since 1981, important changes have occurred in research methodology.

Table 2. Problems of some legume crops.

Crops	Problems
Alfalfa	
- Area 1987: 62300 ha	- Seeds are generally of low quality: local or uncertified, untreated, uncleaned
- Mean yield: 44 tons fresh matter/ha	- Rough soil preparation: not deep enough for lucerne, rarely leveled
- Expected yield: 100 tons fresh matter/ha	- Broadcast sowing - Weed and pest control rare
- Main use: cut as green fodder, hay	
Berseem - Area 1987: 49400 ha - Mean yield: 45 tons fresh matter/ha - Expected yield: 100 tons fresh matter/ha - Main use: cut as green fodder	- Late sowing for berseem - Wrong chemical fertilization: Nitrogen usually applied while phosphorus and potassium are almost never applied - Recommended harvest schedules not respected: cuttings are done at early stage of maturity→low persistence
Vetch (in mixture with oat)	
- Area 1987: 62300 ha - Mean yield: 18 tons fresh matter/ha - Expected yield: 30 tons fresh matter/ha - Main use: hay	- Broadcast sowing - Wrong seed rate: the recommended seed rate is 80kg/ha vetch and 40kg/ha oats. The farmers use 30–50 kg/ha vetch and 50–60 kg/ha oats→low quality mixture - Seeds of low quality - Weed and pest control non- existent
Faba Bean	
- Area 1987: 210900 ha - Mean yield: 750kg/ha - expected yield: 2000kg/ha - Main use: animal and human consumption	- Seeds of low quality - Pest control rare: *Orobanche* and *Botrytis fabae*cause severe damage - Wrong fertilization

The National Institute of Agronomic Research (INRA) adopted a new approach and gave greater importance to on-farm research. The experiments combine small plots with commercial field testing. This approach has given encouraging results, especially with the ley-farming system and winter chickpeas.

Current Developments

Recently there has been increasing interest in the introduction of legume pastures to replace grazed fallow. After considering the success of local experiments, the Ministry of Agriculture and Agrarian Reform decided in Spring 1985 to start the 'Ley-farming operation'. Consequently, about 20 000 ha were sown during 1985/86: 18 500 ha Medic and 1500 ha subclover.

Table 3. The on-going work on legumes.

Organizations	Theme	Species	Objectives	Scientists
National Institute of Agronomic Research	Introduction and comparison of species and cultivars	Medic, subclover, sulla, lupins, strawberry clover, Ladino clover, serradella, berseem persian clover, vetch, pea	– Replacement of grazed fallows by temporary pastures – Reconversion of marginal cropland into pastures	M. Bounejmate G. Jaritz T. Schulte E. Mzouri M. Mazhar C. Al Faiz P. Beale
	Plant breeding	Medic	Select cultivars for – Disease resistance – Hard-seededness – Earliness	M. Bounejmate C. Al Faiz M. Derkaoui
		Chickpea	– Improve winter chickpea	M. Kamel
		Vetch	Select cultivars for – Disease resistance – Cold tolerance – Growth habit	C. Al Faiz
		Lentil	– Improvement and protection	M. Solh
Agronomic and Veterinary Institute Hassan II	Plant breeding	Alfalfa	– Evaluate local strains – Develop synthetic cultivars for oasis and irrigated areas from local and imported material	A. Birouk
	Nitrogen fixation	Faba bean, sulla	– Evaluate strains on productivity	M. Hilali
Faculty of Science, University Sidi Mohamed Ben Abdallah, Meknes	Nitrogen fixation	Annual medic	– Response to inoculation	M. Ismaili

The outstanding success obtained during the first year encouraged those responsible to continue their efforts. 14 000 ha and 7000 ha were sown in 1986/87 and 1987/88 respectively.

After three years experience, interest in the conversion of grazed fallows to annual legume pastures has increased substantially. It is planned to sow 12 000 ha next year.

Reference

Amine, M. and El Adlouni, F. 1987. Production des legumineuses alimentaires au Maroc. Journees d'etudes, Settat, Avril 1987.

Discussion

Falcinelli:
Our experience at Perugia University showed that it is possible to improve alfalfa persistence under frequent cutting regimes. You can try the same in Morocco to improve your local types for frequent cutting regimes with an appropriate breeding program.

Bounejmate:
Yes, we can, but we have also to persuade farmers to follow the recommended schedules. The cultivars actually used (African, Sonora and Moapa) are well-adapted to frequent cuttings.

Cocks:
What has been the impact of the Moroccan project to introduce ley farming?

Bounejmate:
The 7 years' experience and the methodology used helped to make the introduction of the 'Operation ley-farming', 1985–86, easier. Local data was available, and recommendations adapted to local conditions could be made. The cooperation with farmers directly interested in livestock production established a working ley-farming system which served as a demonstration for neighboring farmers and helped to persuade decision makers. The close cooperation with extension people from the beginning was valuable.

Webber:
In the years when large areas of medics were sown and grew successfully, has seed production been part of the program, particularly on individual farms?

Bounejmate:
Right from the beginning of 'Operation ley-farming' a complete plan was made, including technician training, the use of television and radio, field days, and local seed production of medics. In 1985–86, 500 ha were sown for seed production and a decision taken to import vacuum harvesters. Now 4 machines (Bagshaw) are available and seed production has started. Last year, in Sogeta, 30 ha were harvested, with a yield of about 140 kg/ha for Cyprus and 220 kg/ha for Snail (Total production 9800 kg).

Gintzburger:
How did you get farmers to manage the medic – cereal system properly, especially the medic phase?

Bounejmate:
It is not easy, and after 3 years some farmers are still making mistakes, such as grazing late, leading to weed infestation; getting the carrying capacity wrong; and cutting or grazing during flowering the first year. In spite of this, many pastures are well managed, due to the efforts made to train technicians and farmers.

The Role of Legumes in the Farming Systems of Portugal

A.M. DORDIO

Forage Section, Estacao Agronomica Nacional, 2780 Oeiras, Portugal

Abstract. Legume production occupies about 9% of the total area of the country. Since 1963 the areas and production of both feed and food legumes have greatly decreased as a result of high production costs, poor farming methods requiring a large labour force, and low yields. It is important to increase feed legume production, faba beans, peas and sweet lupins, for the mixed feeds industry, to replace in part the large imports of proteinaceous raw materials. Human diets could be improved by increased consumption of chickpeas and beans.

Pasture and forage-growing areas could be increased to 30% of the country if suitable farming systems were used. Legume crops will help to improve soil fertility through the process of symbiotic nitrogen fixation, and should always be included in farming systems on poor soils. They can contribute to higher yields if used in rotations with cereals on better soils.

Constraints to productivity are lack of suitable cultivars, poor farming techniques and little mechanization, and some of the main research priorities aim at overcoming these constraints. National Research Stations and the Regional Directorates for Agriculture are largely responsible for research and extension.

Introduction

Legume production in Portugal occupies 800 thousand hectares, about 9% of the total area, of which 5.9% consists of pastures and forages, 2.8% of feed and food legumes, and 0.3% of horticultural legumes.

Pasture and Forage Legumes

Table 1 shows the areas with temporary and permanent pastures on dryland and irrigated land, including winter forages, according to the latest evaluation in 1982. It also shows the areas where in an improved farming system the recommended crops should be grown. Temporary and permanent dryland pastures are composed of the legumes subterranean and crimson clovers, annual medics, serradella, white and strawberry clovers, and birdsfoot trefoil; and some drought-resistant perennial grasses, fescues and cocksfoot. Irrigated temporary pastures include red, white and strawberry

A.E. Osman et al. (eds.), The Role of Legumes in the Farming Systems of the Mediterranean Areas, 93–103.

Table 1. Forage and pasture acreage (1000 ha); actual and projected.

	Winter	Dryland pastures		Irrigated pastures	
	forages	Temp.	Perm.	Temp.	Perm.
Actual	327	15	169	10	44
Projected	448	1142	956	92	34

Source: Ministerio da Agricultura (1982).

clovers and alfalfa, and hybrid and perennial ryegrasses, fescues and cocksfoot: permanent pastures under irrigation are based on white and strawberry clovers, and ryegrass, fescues, cocksfoot and velvetgrass.

Winter forages are normally composed of vetches, chickling, yellow lupin, sweet clover, Persian clover and berseem, mixed with oats, barley, rye and Italian ryegrass.

Areas with forages and pasture could be increased from 565 thousand hectares to more than 2.7 million hectares, about 30% of the whole area of the country. During the past six years in fact, there has already been an expansion of these crops in a program aimed at developing livestock production.

Food and Feed Legumes

Table 2 shows the areas and yields of faba bean grown for feed, and beans and chickpeas for human consumption. There has been a decrease in total areas and yields here too. Productivity of the three crops has risen slightly, however. The reduction in area is due to the increase in costs, poor cultural practices and the low-yielding varieties used.

Consumption and Trade

There was a continuous reduction in the yield and consumption of faba beans until 1983. However, there were always some to export (Table 3). They used to be grown as feed for the draught animals, horses, mules and

Table 2. Area (1000 ha), production (1000 t) and yield (kg/ha) of feed[1] and food legumes.

Year	Faba bean			Chickpea			Beans		
	Area	Prod.	Yield	Area	Prod.	Yield	Area	Prod.	Yield
1963	76	41	533	70	26	374	426	60	141
1973	43	26	602	39	15	394	306	50	165
1976	42	26	626	41	15	357	270	32	118
1980	35	21	604	40	15	368	273	42	153
1983	28	16	569	29	8	287	228	35	155
1986	23	17	760	25	12	502	198	44	224

[1] Faba bean

Source: Instituto Nacional de Estatistica (1963–86).

Table 3. Production and utilization (1000 t) of food and feed legumes 1976–86.

Year	Production	Seed	Export	Import	Consumption
Faba beans					
1976	26	3	1.00	0.33	22
1980	21	4	0.64	0.36	17
1983	16	3	0.64	0.30	12
1986	17	3	2.00	0.77	13
Chickpeas					
1976	15	2	0.36	0.10	12
1980	15	2	0.65	0.19	12
1983	08	2	0.11	4.00	11
1986	12	2	0.34	2.00	13
Beans					
1976	32	7	3.00	12.00	35
1980	42	6	1.00	9.00	43
1983	39	5	1.00	2.00	35
1986	44	5	0.57	5.00	44

oxen, before mechanization. With the development of the mixed feed industry in the sixties demand for faba bean fell tremendously.

Chickpea consumption also declined until 1976, then stabilized. Production was sometimes inadequate, however, and in 1985 the equivalent of 20% of local production had to be imported. The situation with beans consumption was much the same, except that it rose slightly after 1976.

In 1985, horticultural legumes, beans, peas, and faba beans, grown for canning or as fresh vegetables, occupied an estimated 29 thousand hectares; 54% green beans, 25% peas and 21% faba beans. There has been a substantial increase of pea and faba bean consumption over the last few years (Palma 1987).

Physical Environment

Located in the extreme west of the Iberian peninsula, between the latitudes 36° and 42°, Portugal is approximately rectangular in shape, 160 km wide and 560 km long, with an area of nearly nine million hectares. More than half of the area north of the Tagus river consists of a complex relief system along a narrow and varied coastal strip separating the sea from the foot-hills of the Iberian Plateau (400 mm). Inland the hills rise to more than 700 m, reaching a maximum altitude of 2000 m in the Estrela Mountains. South of the Tagus the relief is less pronounced and only 3% rises above 400 m.

Climate

Rainfall in the north averages 800 mm, increasing to 1200 mm in the north-western corner and 2500 mm in the highest altitudes. In the south average rainfall is between 700 and 900 mm in the coastal zones, and 450 to 700 mm inland. It is concentrated between October and April, and is scarce in the summer, when irrigation is required, even in the areas of high average rainfall.

Average minimum and maximum temperatures in the south vary from 3.5°C to 33°C inland, and from 8°C to 26°C in the coastal areas. In the north they range from 2.5°C to 33°C inland, and 5.6°C to 24°C on the coast. In the mountains the average minimum may be substantially lower in winter, when there is snow.

The average growing period is limited by drought, and varies from seven to eight months in southern inland areas to nine months in some zones on the coastal strip. In the north-west it can exceed nine months.

Although Portugal is not exactly in the Mediterranean Basin its climate is characteristically Mediterranean, with the exception of the north-western corner which is more temperate.

Soils

Portuguese soil is mostly poor, thin and stony, with low fertility, often as a result of bad management. North of the Tagus the soils vary from granitics to shales. In the south, soils derived from shales – most of them skeletal – and sandy soils predominate. There are some small areas with clay soils from diorites, basalts and limestones, in the western central area, for example. They have pH levels varying between 4.5 and 5.5 in the north, and 5.6 and 6.5 in the south. Some exceptional soils derived from limestone and basic eruptive rocks have a pH of over 8.

Land Use

Only 28% of the total area is suitable for farming. The rest should be forest, pasture and other non-cultivated land. At present 55% of the land is farmed. This misuse of 27% of the land is responsible for soil erosion and degradation, and for the low production of several of the crops, especially cereals, which occupy the largest part of this area. The government subsidizes cereal prices so that these farms can remain economically viable. This accounts for the farmers' lack of interest in changing to other crops.

In 1992, as a full member of the EEC, Portugal has to put a stop to the subsidies and adopt Community prices, so most of the less productive farms will have to stop growing cereals. Others, with more fertile soil, should integrate cereal crops in well-balanced farming systems and adopt a technology which will allow higher productivity at lower costs, making them competitive with other EEC countries.

In these circumstances, legumes, as plants for soil improvement, should play a major role, especially as temporary and permanent pastures on infertile soils, and as forage, feed and food legumes in rotations with cereals on the more fertile soils.

Present Role of Legumes in Farming Systems

Pasture and Forage Legumes

Pastures with legumes and forage cultivation on a small scale should be increased as indicated in Table 1, to improve land utilization, particularly in marginal areas. This will increase feed availability and in turn ruminant production, especially of sheep and goats, for local consumption and export. The legumes will contribute towards soil fertility through nitrogen fixation, when appropriately nodulated; through their deep root systems that will utilize and recycle nutrients at depth; through residues that increase organic matter and improve soil structure; and through protection of the soil from erosion. (Almeida 1980; Crespo 1980; Reeves 1984).

In addition, legumes produce more green matter than grasses without nitrogen fertilization; have higher protein and calcium content; have the same digestibility; and voluntary intake is higher (Crespo 1983).

Food and Feed Legumes

The production of feed legumes in Portugal should be greatly increased to reduce the large imports of oil seed and meal which are used as proteinaceous raw materials in the mixed feed industry. The only oil seed crop produced in the country is rainfed sunflower, with a production of nearly 33 thousand tonnes in 1986, which amounted to only a quarter of the imports of this seed. Increased cultivation of oil seed crops other than sunflower, grown mainly under irrigation, may be enough to avoid imports, and may include soya bean if there are benefits from EEC subsidies and if it can compete in foreign markets.

The most widespread feed legume has been local varieties of faba bean, with yields of 760 kg/ha in 1986, but other legume crops produced on small areas have higher yields, such as peas (3000 kg/ha) and sweet white lupin (2600 kg/ha). This lupin has the advantage that it grows in acid soils, and the protein content of the seed is >32%, comparing well with the 24–26% of other seeds. This grain is harvested for the mixed feed industry. New varieties of small faba bean and desi chickpea, a winter crop, are also of interest for animal feeding. The prices of faba bean, pea and sweet lupin are now subsidized by the EEC and a rapid growth in production is expected. The annual per capita consumption of food legumes, kabuli chickpeas and beans, is 1.2 and 4.4 kg respectively. It could be increased with advantages for health, replacing animal proteins (CN FAO 1985).

Production costs condition the cultivation of all legumes and will only be reduced by new varieties and new technology.

Crop Rotations

The Regional Directorates have advised farmers to change the traditional rotations in some of the main areas to improve soil management. Legumes

Table 4. Traditional and improved rotations.

Areas	Traditional rotations	Improved rotations[1]
Northwest Irrigated	Italian ryegrass × maize (grain)	maize (silage) – winter forage or legume maize (grain) – temporary pasture (2 yrs)
Central West Irrigated	winter forage × maize (grain) or or Italian ryegrass potatoes	maize – potatoes – temporary pasture (2 yrs) or horticultural crop
Tagus Valley Irrigated	hortic. – wheat – maize – oil seed (tomatoes) hortic. – hortic. – wheat – maize (tomatoes)(melon)	horticulture – wheat – temporary pasture (3 yrs) hortic(tomatoes) – wheat – winter forage × maize(silage) – winter forage × maize(grain)
South Inland		
Dryland (good & medium clay soils)	summer fallow – oil seed – wheat – barley (1–2 yrs) summer fallow – wheat – barley	oil seed – wheat – legume – barley or or legume winter forage
(poor soils)	cereal – winter forage – fallow (7 yrs)	temporary pastures (7 yrs) – winter forage – cereal
		permanent pastures

[1] Some of these rotations are already being followed.

Source: PMA 1982.

were introduced in all rotations where the succeeding crop can take advantage of the legumes' effect on the soil (Table 4). On soils of moderate and good fertility legumes are grown on 25% of the area, and on poor soils on 50% or more. The cycle of the rotation, normally four or more years, was also taken into consideration to avoid diseases and pests specific to certain crops. The timing of seeding and harvesting, the root systems, and the need for herbicides of different selectivity have also been studied (Aleixo 1988).

On poorer soils on lands marginal for cultivation large areas remain as native pastures for very long periods, or they might be cultivated with a cereal crop, oats or triticale, and sometimes with yellow lupin, every seven years. Seeding permanent and temporary pastures is advised. Stocking rate on these pastures could increase from 0.5–1 to 4 ewes/ha or higher. Table 5 shows the results of a grazing trial on subterranean clover pasture conducted for seven years in Elvas (Eastern center), on low/medium fertility soils, where the best liveweight lamb production, 178 kg/ha, was obtained with a stocking rate of 8 ewes/ha (Crespo 1981). In some areas where the legume population in native pastures is high (serradella, clovers, vetches, birdsfoot trefoil etc) it is more profitable to improve native pastures by liming and fertilizing with phosphorus and trace elements, and by improved grazing management, thus reducing tillage and seeding costs (Dordio 1988)

In these areas, especially south of the Tagus, there are often large areas of cork oaks and green oaks which have a beneficial effect on pastures, contributing acorns for animal feeding during the winter (Mendes 1986). It is important to emphasize, however, the nuisance of several species, especially *Cistus*, which are not eaten by animals. They are removed either with a heavy disc harrow or with herbicides, an expensive and often ineffective process.

Mechanization

In Portugal in 1986 according to the latest statistics there were more than 75 thousand tractors (INE 1988) and nearly five thousand combine harvesters (FAO 1987). Since then the numbers have increased due to community assistance programs aimed at modernizing agriculture. The distribution of

Table 5. Lamb and wool production on subterranean clover pasture at different stocking rates.

Stocking rate Ewes/ha	Lamb production			Wool production	
	livewt. kg/ha	livewt. kg/ewe	growth kg/day	kg/ha	kg/ewe
4	111	27.75	0.257	10.99	2.75
8	178	22.25	0.235	19.30	2.41
12	178	14.83	0.218	27.05	2.58

tractors varies according to the size of farm, income and type of crops. In some rich areas there are too many tractors that may be under-utilized; in other poorer areas there are not enough.

If forage and pasture production is to be increased in new areas it will be necessary to have more machines such as mowers, hay conditioners, forage harvesters, balers, and drills for small and large seeds and for minimum tillage. The availability of equipment used for cereal crops should be taken into account: the combines could be adjusted for harvesting and threshing legume seeds by adjusting the concave, replacing the sieves, and hooking up the nozzle to cutting bars for lodging species.

However, a large number of farms are small and divided into fields where ordinary machines can only work with difficulty, notwithstanding the fact that mechanization is necessary.

It seems that the current rentability of a great number of farms could be maintained if a program were created for the production of beef cattle and sheep using pasture, forage and feed legumes, with reduced inputs, and if EEC subsidies for feed legumes and sheep are retained.

Research

The main constraints limiting legume production are:

1. Lack of knowledge about liming and fertilization with macro and trace elements (S, B, Mo etc) and their interaction for each soil type.
2. Diseases (powdery mildew, *Ascochyta* blight, rust, root rot, anthracnose, viruses etc).
3. Pests (cutworm, *Sitona*, aphids, *Colaspiderma*, *Bruchus*, nematodes etc).
4. Moisture stresses (low resistance of plants to wet or drought conditions during the vegetative period).
5. Inadequate knowledge about defining selection criteria of local *Rhizobium* strains, and of the behaviour of selected soils where there exist native populations of *Rhizobium* effective on cultivated plants.
6. Difficulties in seed production.
7. Low yields of local varieties of food and feed legumes.
8. Infestation by *Orobanche* and other weeds.
9. Unsuitable varieties for mechanical harvesting due to short stems, different stages of maturity of pods, and shattering.

Research priorities are aimed at overcoming these constraints by (i) breeding new varieties resistant to pests, diseases and drought, and with greater yield and nutritive value; (ii) selection and ecological study of rhizobia strains for use in marginal conditions and for artificial inoculum; (iii) studying cultural techniques such as tillage, seeding, fertilization and weed control; (iv) research into pasture management including grazing management, soil conservation and improvement of natural pastures; and (v) further experiments with the technology of seed production, irrigation systems, forage conservation and mechanization.

Research into part of these objectives has been in progress for several decades now, and very important results have already been achieved. There are several new varieties of oats, vetches, lathyrus, berseem, Persian and subterranean clovers, and sweet white and yellow lupin from National Agricultural Research Stations now registered in the National Variety Catalog. Pasture species including annual medics and serradella, and the food and feed legumes, chickpeas, faba beans, lentil, pea, and yellow and sweet blue lupin are being evaluated and improved.

There have been several experiments on liming and fertilization, and Rhizobia lines are being analyzed. Improved native pastures, stocking rates and grazing are currently being tested. Seed production technology has been improved for forage, feed and food legume seed.

Two research projects on intensive farming systems in irrigated areas are taking place at the moment in the central part of the country. They are comparing five different systems, studying the adaptation of the crops and doing an economic analysis of the results. They are also measuring changes in the physical and chemical characteristics of the soil in each system, mainly to determine the influence of temporary pastures. The trials began three years ago, and although a large amount of data has been collected, there are no conclusive results yet.

Other studies on farming systems are being carried out at the National Agronomic Research Station, Oeiras, in the Forages and Agronomy Department; and at the Advanced Institute of Agronomy, Lisbon; the University of Evora; and the University of Tras os Montes, Vila Real.

At the Research Stations there is a shortage of staff, equipment and money.

Extension and Technology

At present extension services are the responsibility of seven Regional Directorates for Agriculture. The regions are sub-divided into Agrarian Zones which in turn include several Council Brigades, where extension agents advise farmers. These agents are supported by specialised technicians in the Zones, who in their turn depend on specialists working at the Regional Directorates in contact with Research Centers. Periodically, updating courses for the technicians are held, according to their categories and specializations. Scientists involved in the various areas of research participate in these courses.

Research centers are constantly referred to by regional specialists, and by the more educated farmers. They promote annual Open Days which are well attended. Regional Directorates and technicians collaborate frequently in research projects, together with National Research Stations. Research reports are published in specialized journals, and scientists must also write technical publications, and prepare articles for regional agricultural newspapers, and texts for radio and television broadcasts. Some important agricultural cooperatives and professional associations employ technicians at different levels to assist their members. On-farm research concerning pasture, feed and food legumes has not been well developed. Only some research and numerous observation and demonstration trials have been carried out on farms.

References

Aleixo, A.L. 1988. Rotacoes e tecnicas culturais com referencia as proteaginosas. I Jornadas Portuguesas de Proteaginosas. EAN. Oeiras, Portugal.

Almeida, J.R.M. 1980. Forragens para verde. Documentacao sobre forragens e pastagens. INIA. Lisboa, Portugal.

Associacao Portuguesa dos Industriais de Alimentos Compostos para animais 1987. Relatorio da Direccao. Lisboa, Portugal.

Comissao Nacional da FAO 1985. Previsao das necessidades dos alimentos para a populacao portuguesa. Lisboa, Portugal.

Crespo, D.G. 1980. Pastagens semeadas temporarias e permanentes de sequeiro. Documento sobre forragens e pastagens. INIA. Lisboa, Portugal.

Crespo, D.G., Antunes, J.C., Dias, J.S. 1981. Influencia dos encabecamentos na producao de carne de ovino e la em prados de sequeiro. Pastagens e Forragens I: 90–95.

Crespo, D.G. 1983. Legume production. Proceedings of the 9th General Meeting of European Grassland Federation, Reading 5–9 Sept. 1982. British Grassland Society. Hurley, U.K.

Dordio, A.M., Monteiro, M., O'Neill, J.F. 1988. Melhoramento de pastagens naturais no mioplioceno. Pastagens e Forragens 8, In press.

Food and Agriculture Organization of United Nations, 1987. FAO Production Yearbook Vol 40. Rome, Italy.

Instituto Nacional de Estatistica, 1963, 1973/6, 1980/3/6. Estatisiticas Agricolas – Continente, Acores e Madeira. Lisboa, Portugal.

Mendes, F.L. 1986. A floresta e as pastagens. Terrenos silvopastoris. Montados. Floresta de proteccao. I Congresso Nacional Florestal. Comunicacoes. Lisboa, Portugal.

Ministerio da Agricultura, Comercio e Pescas, 1982. Programa de Mudanca da Agricultura. Lisboa, Portugal.

Palma, M.D. and Tavares, H. 1987. Caracterizacao da producao agricola – Frutas e horticolas. Relatorio. Direccao Geral de Planeamento e Agricultura. Lisboa, Portugal.

Reeves, T.G. 1984. Lupins in crop rotations. Proceedings of 3rd International Lupin Conference. ILA, 4–8 June 1984. La Rochelle, France.

Discussion

Springborg:
What is the difference between the EEC-supported food legume price and the price in Portugal for the past several years?

Dordio:
For the producer the price is more or less the same: for the mixed feed industry the EEC subsidy will reduce the price by 45%.

Capper:
You have suggested that there is considerable potential for feed legumes to replace soya bean meal in mixed feeds. To what extent has the feed manufacturer been involved in this process, as it is easier for them to use a consistent product such as soya bean meal than a large number of different materials of inconsistent quality?

Dordio:
It is largely a matter of price. Feed manufacturers have been using feed legumes, partially replacing soya bean meal, since their price is competitive. They are buying local peas at an EEC-subsidised price at thc moment. Blue lupin seeds are being imported from Australia, without a subsidy.

Buddenhagen:
Please explain how lupins are used – for grazing, hay or grain – especially in relation to grazing under cork oaks.

Dordio:
Yellow lupins are generally grown on poor soil under cork oaks and are used for very light grazing in spring, and summer, when they are dry. In this case the seeds are spread on the ground, and those that are not eaten will generate a new crop. Sweet white lupins, produced in medium-fertility soils, are used for grain as a component of mixed feeds.

Papastylianou:
You have given several suggestions for improving the farming system and introducing more legumes. What are the chances that these suggestions will be adopted?

Dordio:
It is already happening in certain areas. In others, the future decrease in cereal crops, with low yield, will be favourable for the introduction of forage and feed legumes in order to maintain the same profit.

The Role of Legumes in the Farming Systems of Syria

B. MAWLAWI and M.W. TAWIL

Directorate of Agricultural and Scientific Research, Douma, Syrian Arab Republic

Abstract. Syria may be divided into five agricultural regions, termed 'Stability zones', on the basis of average annual rainfall. Roughly 45% of the land is cultivable. Food legumes (chickpea, lentil and faba bean), and forage legumes (*Lathyrus*, *Vicia* spp.), grown in Zones 1 and 2, (250–600 mm) are important in the economy, although the most important feed resource is natural pasture, supplemented with crop residues, barley forage and grain, and some concentrates. There is a serious feed shortage. The most common crop rotation is cereal (wheat, barley) – fallow, although legumes have been introduced into some rotations. Constraints presently limiting legume production and productivity are pests and diseases, low level of inputs, and labour costs. Research strategies are aiming at extending new technologies through on-farm trials, particularly towards an integrated farming system (sheep, medic pastures-cereal); studying stocking rates; and breeding new cultivars of lentil and chickpea which are resistant to diseases and pests, tolerant to cold, and suitable for mechanization.

Introduction

The total area of the Syrian Arab Republic is more than 18 million ha of which 8 million hectares are cultivable land and the remainder desert and rocky mountains. The Syrian desert is used for grazing when there has been sufficient rainfall.

Geographically, Syria may be divided into four regions:
(1) the coastal region (2) the mountainous region (3) the interior region or the plains and (4) the desert region.

Physical Environment

Syria may be divided into four climatic regions, determined by rainfall, which coincide with the four geographic regions. The two to the West, the coastal and mountain regions, have a characteristically Mediterranean climate with heavy rainfall in winter and high relative humidity in summer, and moderate temperatures, decreasing with altitude in the mountains, where it can snow in winter. The two interior regions, the plains and the desert, have less rainfall, in winter only, colder winters and very hot summers.

A.E. Osman et al. (eds.), The Role of Legumes in the Farming Systems of the Mediterranean Areas, 105–114.

Temperature

The daily range between the maximum and minimum temperatures is generally quite high in most of the country. December and January are the coldest months of the year while July and August are the hottest. In winter the temperature frequently falls below 0°C, but rarely gets down to −10°C, while in summer it may frequently rise to 48°C.

Precipitation

The heaviest rain falls in the mountainous and coastal regions, followed by the northern region (North Aleppo, Kamishly and Malikieh). The country is subject to mild droughts from time to time, which may seriously reduce agricultural production. Based on rainfall intensity Syria may be divided into five agricultural zones, termed 'stability zones':

Zone	*Average rainfall (mm)*
1a	>600
1b	350–600
2	250–350
3	250
4	200–250
5	<200

Land use in relation to the stability zones is illustrated in Fig. 1.

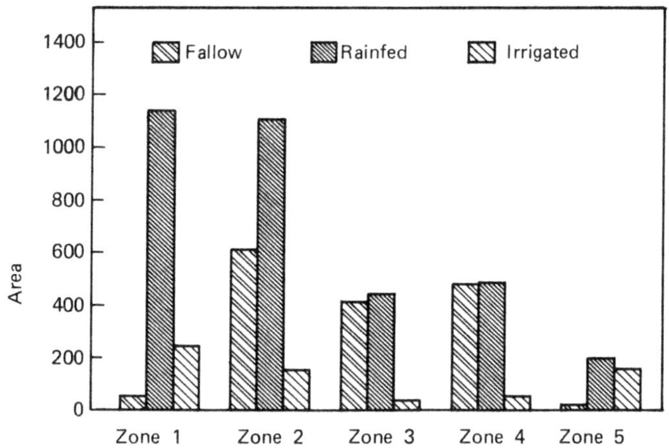

Fig. 1. Cultivated land use (1000 ha) according to stability zones, 1986. *Source*: Annual Agricultural Statistical Abstract 1987. Ministry of Agriculture and Agrarian Reform, Damascus, Syria.

Legume Crops

Forage Legumes

The forage legumes (*Lathyrus* spp., *Vicia* spp.) are cultivated in the first and second zones. They can be included in the crop rotation in areas dominated by cereal cropping systems. They may also be cultivated in irrigated areas in a very limited way. In general the production and area cultivated with these crops increase in wet years and decrease in dry years.

Table 1 shows area, production, and yield of the vetches (*Vicia sativa*, *Vicia ervilia*, *Lathyrus* spp), 1976–1985.

The area cultivated to grazing vetch increased by more than four times with no corresponding increase in yield. Yield was very low in 1985 because of drought, only 1.2 t/ha as green material, in contrast with 1980 when it reached 18.80 t/ha because of the favorable climatic conditions during the growing season. The area of grain vetch decreased from 16 thousand hectares in 1976 to seven thousand hectares in 1985 because of high costs of agricultural practices, especially hand harvesting. The mean yield was 800 kg/ha, but it fluctuated between the seasons. The area of *Vicia ervilia* decreased in the same period from 25 thousand hectares to 17 thousand hectares. The area cultivated to *Lathyrus* dropped to 29 thousand hectares in 1985. The yield fluctuated due to erratic distribution of rainfall.

Pasture Legumes

Natural pastures are the most important feed resource for sheep and goats in Syria. Forage production varies from year to year due to climatic conditions, especially annual rainfall (140 mm average in the Syrian Steppe). The forage productivity of steppe pastures and fallow land is about 27 million tons dry matter reaching a peak during March-June, with dry

Table 1. Area (1000 ha) and production (1000 t) of major grazing and grain legumes in Syria 1976–85 under rainfed conditions.

Year	Grazing vetch		Grain vetch		Bitter vetch		Chickling vetch	
	area	prod.	area	prod.	area	prod.	area	prod.
1976	8	32	16	13	25	15	39	32
1977	2	30	15	10	27	16	38	30
1978	4	30	11	6	25	14	41	28
1979	2	25	10	5	16	7	28	18
1980	2	31	8	7	21	14	33	32
1981	3	28	7	6	16	11	32	21
1982	4	42	6	5	11	7	33	17
1983	3	27	7	8	18	11	33	27
1984	24	25	6	6	12	6	27	19
1985	34	41	7	5	17	9	29	18

Ministry of Agriculture Annual Abstracts (1985).

residues of forage plants lasting until September. This amount of forage production provides about 119,000 t of digestible protein (DP) and 1,465,000 t of total digestible nutrient (TDN), which represents roughly 36% of the total local forage production of each of these two components.

In comparison, cultivated green forage production contributes only 3.5% of total feed production, 264 thousand tonnes as dry matter of which 74.2% is grazed rainfed forage (15% of barley rainfed area and 8% of wheat rainfed area). The forage legume production (*Lathyrus* spp., *Vicia* spp.) around the year is 41 thousand tonnes as dry matter (15.5% of total green forage production). This means there is an acute shortage of green forage production for cattle and sheep. Crop residues (barley, wheat, food legumes) contribute 2.9 million tonnes. Since the straw is the main component of crop residues the DP content (28,300 t) and the TDN (1,273,000 t) are very low. These crop residues are consumed by all livestock. Concentrates rate third after pasture and residues, and the total production is 1,546,000 t dry matter, 165,000 t DP and 1130,000 t TDN. They consist of cereals (63.3%), by-products of agricultural industry (34.4%), and grain legumes (0.3%). Barley grains are the main component of forage cereals (93.8%), followed by maize and sorghum (6.2%). Rainfed cultivation of barley is dependent on climatic conditions, and the production of concentrates is affected by fluctuations in rainfall.

Food Legumes

Food legume crops are also important in the economy of Syria. Chickpea and lentil are among the most important of these crops. They are a relatively cheap protein source that partly meets population needs, a source of foreign currency through exports, and an important part of rotations in rainfed areas. The total area grown to these crops in the different environmental regions is about 185 thousand hectares mainly in the first and second stability zones (250–600 mm annual precipitation) (Table 2). Local

Table 2. Area (1000 ha) and production (1000 t) of chickpea and lentil in Syria 1976–85 under rainfed conditions.

Year	Chickpea		Lentil	
	area	production	area	production
1976	67	50	140	127
1977	41	25	172	110
1978	46	31	132	88
1979	17	10	87	41
1980	91	73	82	80
1981	85	63	70	60
1982	56	37	56	51
1983	94	74	70	59
1984	53	37	58	35
1985	79	50	66	47

Ministry of Agriculture Annual Abstracts (1985).

chickpea is sown in spring (Feb – March) while lentil is sown in winter (Dec – Jan).

Livestock and Nutrition

The total number of animal units (AU = 1 mature cow; 5 mature sheep; 6 mature goats; 0.8 horse) in Syria in 1983 was 3700000. There were 13.3 million sheep (8.724 million in the cooperative sector and 5.567 million privately owned), 767 thousand cows and 1.157 million goats (94.6% mountain goats). There were 50 thousand horses. Poultry numbered 64.4 millions as broilers, and 7.6 millions as layers.

Sheep are important in animal wealth, and the annual changes in numbers are directly connected with feed availability, especially in dry seasons. Sheep are highly dependent on natural pastures and crop residues. Barley forage and grain crops replace the natural pasture during drought years.

The total nutritive requirement in 1982–1983 was 7.99 million tonnes as dry matter, (4.77 million tonnes as TDN, 5 million tonnes as DP). Sheep consumed 56 and 59% of the two components, respectively, while cows consumed 19 and 20%, and poultry 10 and 14%, respectively.

Feed balance (i.e. deficit) 1982/83 was about 704,000 t TDN and 136,000 t DP. This shortage affected livestock productivity. Table 3 illustrates the production, consumption, and balance of feed.

Cropping Sequence

The cropping system varies according to zones, and to various factors, mainly climate, soil, marketing facilities, and prices. The current rotations are as follows:

Table 3. Mean feed production, consumption and balance (1000 t) 1982/83.

Source of Feed	Feed Component		
	DM	CP	TDN
Natural pastures, fallow land	2745	119	1465
Green forage (rainfed & irrig.)	264	20	198
Residues (rainfed & irrig.)	2922	28	1273
Concentrates	1546	165	1130
Total	7477	332	4066
Feed imports	308	41.3	240
Total feed available (prod. + import)	7785	373.3	4306
Total feed requirement (1)	7990	468	4770
Feed balance	−205	−94.7	−464
Self-sufficiency (%)	93.6	70.9	85.2

(1) includes requirements of sheep, cows, goats, horses, camels, buffalo, poultry and fish.

Source: Forage Organisation.

Cereal (wheat or barley) – fallow
Cereal (wheat or barley) – legumes – fallow or summer crop

The following rotations can also be observed sometimes in certain areas:

Cereal (wheat or barley) – legumes – summer crop
Cereal (wheat or barley) – legumes
Cereal (wheat or barley) – Rainfed cotton or summer crop
Cereal (wheat or barley) – Tobacco or vegetables

The fallow system is generally dominant (Table 4). (Total fallow area in 1986 was 1.6 million ha).

Land in Syria is fallowed for the following reasons:

1. To improve soil fertility by increasing mineral content, especially nitrogen.
2. To conserve moisture in the soil.
3. To control weeds by ploughing them in fallow land.
4. To control diseases

Problems with Crop Rotations

Rainfall varies according to zones, and to years, and usually fails to meet crop requirements. This variability is reflected in the stability of crop rotations, especially in semi-arid to arid lands, and this is consequently reflected in fluctuating production.

The farmers do not adhere to the government-planned rotation because of economic, social, and cultural factors, such as immigration to the cities, labour shortages and high costs.

Table 4. Planned and actual rotations according to stability zones, 1983.

Stability zones	Planned rotations	% intensification	Actual dominant rotations
1	wheat 50% – legumes 25% – summer crop 25%	92.6	wheat 50% – fallow (or legume/fallow) 50%
2	wheat or barley 50% – legumes 25% – summer crop or fallow 25%	60.3	wheat or barley 50% – fallow 50%
3	barley 50% – fallow 50%	50	wheat or barley 60.5% – fallow – fallow
4	barley 33% – fallow 67%	33	wheat or barley 60.5% – fallow-fallow
5	uncultivated[1]	–	

[1] There are many violations in rainy years, when farmers resort to damaging rotations (barley-fallow).

Mechanization

Tractors and disc harrows are used in land preparation, seed drills in sowing and fertilizer application. Mowers and pellet-making machines are not used because they are generally unavailable, and holdings are small .

Seed and fertilizer are often broadcast by hand, while seed covering is done by disc harrows. Harvesting and threshing are done by hand.

Constraints which Limit Legume Production

The plan is to cultivate forage legumes in order to reduce the percentage of fallow in zones 1 and 2, to transform the traditional rotation from cereal-fallow to cereal-legumes to fill the forage gap, and to diminish pressure on the steppe, especially in the dry seasons where sheep move to rural areas, and sheep owners hire some of the area cultivated to cereal and forage legumes to feed their sheep. This goal has rarely been achieved because of socio-economic considerations, mainly:

1. There is no plan to cultivate forage legumes in livestock areas.
2. There is no complete integration system between livestock and crop production.
3. Small holdings prohibit use of machines on a large scale.
4. High costs of labour.
5. Unavailability of mowers, pellet machines, harvesting machines etc.
6. Pricing policy.
7. Marketing problems, especially for hay, because of the distance of forage areas from livestock areas (large volumes of hay and transportation costs).

Constraints Limiting Productivity of Legumes

1. There is no application of fertilizer, and if it were used it would probably not be used as recommended.
2. Diseases: downy mildew, powdery mildew, bacterial blight, ascochyta blight.
3. Insects: *Sitona* spp., *Bruchus* spp., leaf miner.
4. Parasites: *Orobanche*, nematode.

In addition to the above, chickpea and lentil face the following constraints.

1. Local cultivars are not suitable for mechanical harvesting because of short stems and lodging.
2. Productivity is unstable because the area cultivated each year depends on rainfall, and yields of local cultivars are variable.
3. As wages have risen sharply, mechanical harvesting is becoming obligatory, especially in large cultivated areas where farmers face great difficulty in paying for labour to harvest legumes by hand.

Legume Research

The Objectives and Priorities

1. To breed non-shattering varieties with high yields of seed and biomass, resistant to diseases, parasites and pests, and well-adapted to local conditions.
2. To grow new medic varieties well-adapted to zone 3.
3. To breed lentil and chickpea cultivars suitable for mechanical harvesting.
4. To breed chickpea cultivars that can be successfully sown in winter.

Future Strategies

1. To conduct on-farm trials in order to arrive at an integrated productive farming system (pastures, Medic-sheep-cereal) in zones 1, 2, and 3.
2. To study the stocking rate and its relation to rainfall and soil fertility.
3. To study the extent of the effect of downy and powdery mildews on productivity, and to find highly resistant varieties.

Research Achievements

Extensive high level programmes were initiated at the end of the seventies, with technical and financial cooperation between the Directorate of Agricultural and Scientific Research in Syria (ARC) and ICARDA, to solve some of the problems of legume cultivation. Tens of introduced winter chickpea genotypes were tested to evaluate productivity, adaptability to local environmental conditions, resistance to ascochyta and suitability for mechanical harvesting. Many genotypes were identified. In kabuli chickpea Line ILC 3279 was selected and found to be tolerant to winter cold, resistant to ascochyta blight, high yielding and suitable for mechanical harvesting. It was released by the committee of the National Variety Release in 1986 under the name Ghab-2. The chickpea Line ILC 482 was also selected for its high productivity, relative tolerance to winter cold, and suitability for harvesting mechanically, and was released in 1986 under the name Ghab-1 (Table 5).

The lentil Line 78S 26002 was selected for its high productivity, and suitability for harvesting mechanically. It was released in 1987 under the name Idlib-1 (Table 6).

As the research programme continues, it is anticipated that new varieties will be in the pipeline with higher productivity and suitability for mechanical harvesting.

Extension of Technologies

Agricultural extension has not been playing its role effectively in connecting the researchers and the farmers. Its role was limited to publishing extension

Table 5. Comparison of yield (kg/ha) of local spring chickpea with winter varieties (Ghab-1 and Ghab-2), 1985 to 1987.

Variety	Location	Zone*	Area (1000 ha)	Yield
		1985/86		
Local	Idleb	1	4	1560
Local	Aleppo	1	31	1550
Ghab-1	Aleppo	2	2	2420
Ghab-1	Afrin	1	1.2	3000
Ghab-2	Aleppo	2	2	1740
Ghab-2	Afrin	1	1.2	2400
		1986/87		
Local	Maharda	1	2	1200
Local	Afrin	1	2	940
Ghab-1	Maharda	1	2	3413
Ghab-1	Afrin	1	2	1725
Ghab-2	Maharda	1	2	2115
Ghab-2	Afrin	1	2	1771

*Zone 1 Areas with annual rainfall above 350 mm.
*Zone 2 Areas with annual rainfall between 250–350 mm.

Source: SMAAR/ICARDA Collaborative Program Reports 1985–87.

Table 6. Average production (kg/h) of Idlib-1 in on-farm trials under rainfed conditions.

Season/variety	Idlib-1	Local
1982/83	1110	992
1983/84	1058	737
1984/85	1053	1044
1985/86*	1587	1074

* Results of large scale area (0.5 ha).

bulletins. Nowadays agricultural extension is functioning a little more effectively after establishing extension units at the village level. Some extension is carried out by the Directorate of Agricultural and Scientific Research (ARC) by inviting farmers and technicians to field days. A project started in 1984 to carry out on-farm trials on large areas, 1–5 ha for each field, to assess the possibility of extending medic in zones 1 and 2 through introducing it in the crop rotations to replace fallow (alternating with wheat or barley). These on-farm trials were very successful and very efficient in studying the constraints to applying this farming system. It also helped to solve those problems practically, and had an important extension role by helping in the dissemination of medic cultivation. This has led to a high demand for seed by farmers.

References

Annual Abstracts 1981–86. Central Bureau of Statistics, Damascus, Syria.
Annual Agricultural Abstract 1985. MAAR, Damascus, Syria.

Discussion

Kamel:
There was no mention of faba bean in Syria. Is it less important than lentil and chickpea, or is the improvement of this crop still at an early stage?

Tawil:
Faba bean is more difficult to breed because it is cross-pollinated and needs complicated facilities for isolation etc. There have been no important results so far.

Halila:
How is the suitability for mechanical harvesting expressed in Idleb I? Is it in its growth habit and erectness? If so, does it keep this erectness under very favorable conditions?

Tawil:
Yes. It is suitable for harvesting by cutter bar because of its good standing ability. This erectness disappears in very favorable conditions – 500 – 600 mm rainfall.

Ibrahim:
Research at ICARDA on lodging resistance in lentil genotypes showed that there is a genotype x environment interaction. The two factors studied, plant population and soil moisture, were found to encourage lodging at their highest level. The genotype ILL 8 (released as Idleb I in Syria) would lodge under high population – high moisture conditions.

Guler:
You have concentrated on breeding. Is your only problem lack of suitable varieties?

Tawil:
No, but we think it is the main one.

Kamel:
You have used the term 'zone'. I think we should be very careful when we use this word and always give the rainfall figures as well, since 'Zone A' in Syria might be very different from 'Zone A' in Iraq, for example. In some countries the criteria may include depth of soil, slope, presence of stones etc., as in Algeria. The way the chickpea is eaten in the different countries determines the acceptability of the variety to be released. In the Mashreq countries in general the chickpea is used for hommos, in which case the seed size is not important. In the Maghreb, however, only a variety with a large, attractive seed will be acceptable.

The Role of Legumes in the Farming Systems of Tunisia

M.H. HALILA,[1] A.B.K. DAHMANE[2] and H. SEKLANI[1]

[1]*INRAT, 2080 Ariana, Tunisia* [2]*INAT, 43 Avenue Charles Nicolle, Tunis, Tunisia*

Abstract. Only 10% of the area suitable for growing legumes in Tunisia, i.e. the northern and higher ground in the central zone, is used for legume cultivation. There are both agronomic and socioeconomic reasons for this. Food legume yields are low and variable, with faba bean and chickpea the main crops grown. Better agronomic practices and more suitable varieties would increase yields of food legumes. There is a greater acreage of forage legumes, a vetch and oats mixture being popular and profitable. Annual medics was introduced to replace fallow in the 1970s, and *M. truncatula* cv Jemalong and *M. littoralis* cv Harbinger were selected as the best adapted to Tunisian conditions, but the new system has not been successful, because of technical shortcomings and conservative farmers. Forage legumes, including medics, have been studied to assess their contribution to soil nitrogen, their effect on soil moisture compared with fallow, and their impact on agricultural production. There is no doubt that integration of cereals with livestock should remain the aim of medic/cereal-based rotations.

Introduction

Tunisia has a Mediterranean climate, characterized by hot, dry summers, rainy winters and great variability in weather and rainfall patterns from year to year. Variability in rainfall occurs between seasons but also within a single location. The physiography of the country is also varied.

Climatic patterns along with topography determine three main zones (Fig. 1) each of which has different weather and agricultural activities. These zones are:

- Northern zone; 350 – >700 mm rainfall. Mountainous.
- Central zone; 200 – 350 mm rainfall. High tablelands and pastures.
- Southern zone; <200 mm rainfall. Desert.

The legumes are grown primarily in the northern zone and the higher areas of the central zone.

Legumes in the Farming Systems

Legumes play a major role in agricultural production. A cheap and important source of protein for animal and human consumption, they provide

A.E. Osman et al. (eds.), The Role of Legumes in the Farming Systems of the Mediterranean Areas, 115–129.

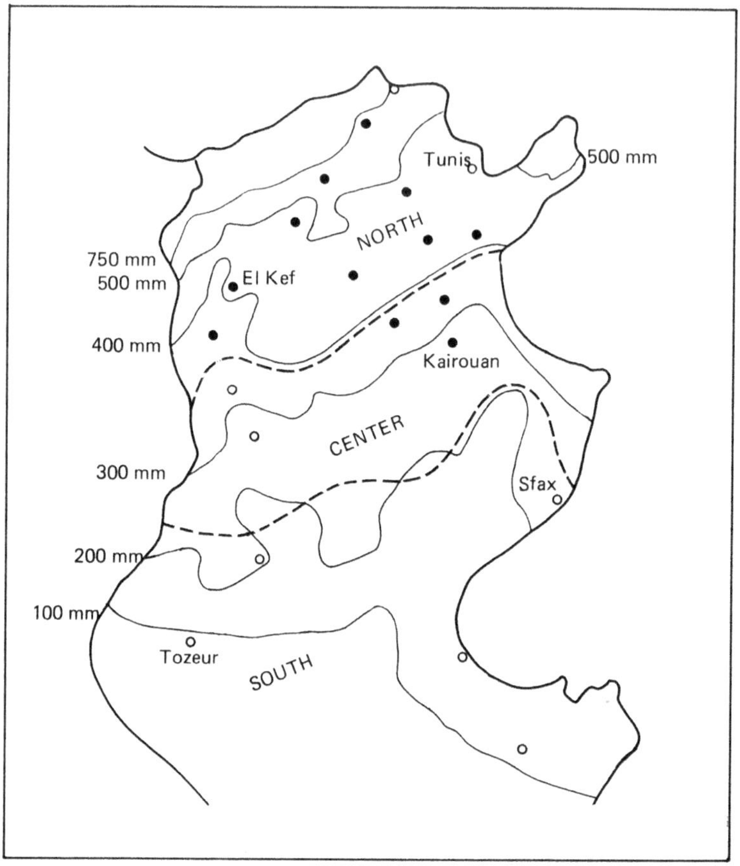

Fig. 1. Bioclimatic map of Tunisia.

rich crop residues as animal feed, and play a key role in maintaining the soil fertility through biological nitrogen fixation. They are present in practically all types of rotations used by farmers (Table 1).

Their role in the economic return and social activities at farm and community levels is undeniable. Their impact on the canning and other industries is also important. However, despite all the benefits attributed to legumes we need to ask: are legumes given enough attention by the farmers and by the agricultural policy makers? The answer is given by the statistics: they are not. Despite all the talk and the written reports in this country about the beneficial roles played by legumes in the various farming systems, they still lag behind, with only an average of 199 thousand hectares per year, (10%), of the total cultivated area in the north (Table 2), while fallow occupies around 400 thousand hectares (21%) every year (Dahmane 1987).

Table 1. Crop rotations used in Northern Tunisia.

Type of rotation	Sequence			
	Year$_1$	Year$_2$	Year$_3$	Year$_4$
'Modified' 2-year	Cereal	Fallow or forage or food legumes or medic		
3-year	Cereal	Secondary cereal or forage or food legumes or flax	Food legumes or fallow or sunflower or sugar beet or tobacco	
	Cereal	Sulla*	Sulla	
4-year	Cereal	Secondary cereal or food legumes	Forage	Food legumes or sugar beet
	Cereal	Artichoke	Artichoke	Food legumes and vegetables
	Cereal	Vetch or corn	Forage and vegetables	Winter and summer vegetables

* *Hedysarum* spp.

Table 2. Average total areas of legumes and corresponding percentages relative to total cultivated area in northern Tunisia (1000 ha).

Legume	Average acreage	Proportion of total cultivated area (%)
Food legumes[1]	86	4.5
Forage legumes[2]	108	5.6
Annual medicago[3]	5	0.3
Total area of legumes	198	10.4
Total cultivated area in the north	1913	

[1] Average of the past 15 years.
[2] Average of the last 6 years.
[3] Average of the last 10 years.

Many different reasons, qualified as constraints to the productivity of legumes, have been proposed to explain this discrepancy.

1. *Agronomic*

Lack of high yielding, stable and good nitrogen-fixing varieties.

Scarcity of agronomic information about practices and biological nitrogen fixation

Insufficient high quality certified seed for grain legumes
Mechanization constraints
Parasite infestation

2. *Economic*
No clear encouraging price policy
Inadequate marketing system
3. *Social and Technical*
Lack of intensive livestock husbandry
Inadequate integration between livestock and other crop production systems
Absence of an appropriate extension network.

Medicago forages are presented in a separate group because of their role in Tunisia with the medic ley system.

Importance of the Food Legumes

Food legumes are certainly the farmer's 'heartiest' legume crops in his agricultural activities and systems. They are, however, second to the forage legumes in terms of acreage (Table 2). Statistics on the acreage (Halila 1986) show tremendous yearly fluctuations with no clear trend (Table 3). Faba bean and chickpea are the main food legume crops, followed by dry pea, lentil and *Phaseolus*. Of course, this fluctuation reflects the season's growing conditions but economic factors, specially product prices, also determine the area to be planted of each crop.

Average yields for all crops are low and are characterized by wide fluctuations (Table 3). This unstable yield is obviously linked to the growing conditions of the season, but is also associated with the lack of adapted stable varieties and the use of inadequate agronomic practices. The disease susceptibility and stress sensitivity of the varieties and populations grown by the farmers are among the major reasons for the instability of yield.

Forage Legumes

Table 4 shows the annual area for each forage legume crop. With the exception of the 1987/88 season, the average total acreage has been more or less stable for the last six years. The individual acreage for each crop is also much the same except for the mixture vetch-barley (V.B) and the fenugreek; both showed wide fluctuation.

The mixture of vetch and oats (V.O) is a popular forage grown widely and consistently with a large acreage. This forage is as profitable as any other cash crop especially during dry years, and during seasons when there is a severe shortage of animal feed it can be more expensive than the subsidized bread price.

Table 3. Fluctuation of the average area (1000 ha) and yield (t/ha) of the main food legume crops during the last 15 years.

Yield	Faba bean		Chickpea		Lentil		Dry pea		Bean	
	area	yield	area	yield	area	yield	area	yield	area	yield
Max	61	1.00	38	1.10	7.0	1.06	9.5	0.86	3.5	1.01
Min	30	0.42	20	0.43	0.5	0.43	5.0	0.37	1.0	0.57
Average	48	0.82	28	0.79	2.0	0.67	7.0	0.79	2.0	0.75
Trend*	V	V	I	D	D	V	V	I	D	D

* V = variable; I = increasing; D = decreasing.

Table 4. Fluctuation of the average area (1000 ha) of the major forage legumes 1983/88.

Acreage	Berseem	Sulla	Lucerne	V.O mixture[1]	V.B mixture[2]	Fenu-greek	Total
Max	5	14	14	139	9	17	188
Min	4	5	8	103	0.2	8	138
Average	4	11	11	128	5	12	171
Trend*	S	V	V	I	V	D	

* S = stable; V = variable; I = increasing; D = decreasing.
[1] mixture of vetch and oats.
[2] mixture of vetch and barley.

Annual Medic

This forage was first introduced in Tunisia in 1971/72 when the cereal project started a research and extension project on annual medic with the objective of replacing the fallow by medic in zones where the rotation fallow-cereals is practised. The concept of integrating livestock to cereal production, thus improving the economic return of the farmers, was quickly implemented and has generated a lot of excitement among farmers, technicians and officials. However, curiously enough, the ley system did not really take off. Statistics for the last 15 years indicate that the areas covered by the medic remain low and fluctuating (Table 5). The largest acreage sown to this forage was achieved in the 1985/86 season with slightly more than 11 thousand hectares which is 12% of what was originally planned in the Sixth Tunisian Development Plan. The actual areas planted with medic represent only 2 to 8% of the area programmed for the same period (Table 6). Many reasons were given to explain the setback to the system (Haddad 1982; Dahmane 1987) as follows:

Table 5. Area (ha) sown to annual medic since 1971/72.

Season	In rotation with cereals	Pasture improvement	Regeneration	Total
1971/72	15			15
1972/73	468			468
1973/74	492			492
1974/75	2665	115		2770
1975/76	2525	602		3128
1976/77	2200	2350	500	5105
1977/78	2691	400	1366	4457
1978/79	1780	1641	896	4312
1979/80	1563	641	1234	3438
1980/81	905	508	2215	3628
1981/82	954	271	2227	3452
1982/83	1243	211	215	1669
1983/84	1030	400	1445	2875
1984/85	3056	1671	3950	8677
1985/86	6665	4802	NA	11467
1986/87	4368	2119	500	6987
1987/88	2564	640	4680	7574

Table 6. Planned and actual total medic area (ha) at the end of the Sixth Development Plan (1982–1986).

Year	Planned		Actual	
	Yearly	Cumulative	Actual	Cumulative (%)
1982	14 000	14 000	1225	8
1983	25 000	39 000	1454	3
1984	25 000	64 000	1430	2
1985	16 000	80 000	4727	5
1986	10 000	90 000	11467	12

- Unavailability and cost of seed
- Lack of adequate equipment for medic ley implementation.
- Lack of phosphate application on medics.
- Inadequate grazing of medic by the animals
- Shortage of well trained extension technicians in the medic ley-wheat rotation
- Inexperience of local farmers with managing livestock
- Attachment of farmers to traditional cropping practices, especially fallowing and deep plowing
- Delicate and tricky system
- Prevailing fallow lease system

However, the system itself was never blamed. On the contrary both authors believe that the system is technically feasible and socially possible.

Analyses of the acreage data for the last ten years show that the main feature preventing the establishment of the system is the loss of the newly seeded medic area. Indeed, during 1978–88 more than 70% of the medic sown during that period was lost and did not regenerate in 1988.

In our opinion, the reasons listed above are each valid to a certain extent. We believe, however, that the concept of integrating livestock production with cereal production was rarely digested by most of the farmers who tried the ley farming system. Moreover, the system was first targeted to areas where fallow, 'real', weedy, or any other fallow type, has been common practice since agriculture started: fallowing is deeply anchored in the agricultural system.

The Future Potential of Legume Crops

Food Legumes

Research data obtained during the last few years (Laboratoire des Legumineuses Alimentaires 1986) show the potential of the food legumes and their beneficial impact in improving the economic return of the farmers. Two examples illustrating this potential are presented in Tables 7 and 8. The first is the improvement of yield of local chickpea and lentil cultivars by

Table 7. Comparison of yield (kg/ha) of winter (W) versus spring (Sp) sown chickpea using three chickpea genotypes,[1] and their relative yield increase (%) over the national average (NA).[2]

Genotype	W	Sp	W/Sp (%)	W/NA (%)	Sp/NA (%)
Chitoui	2056	1145	79	160	44
Kassab	2714	1470	84	243	86
Local	2010	1339	50	154	69

[1] Average of 3 sites during 1985–87.
[2] National average yield for the last 15 years.

Table 8. Average yields (kg/ha) of some lentil genotypes in relation to the national average (NA).[1]

Genotype	Average Yield[2]	Yield/NA (%)
Nsir	1606	139
Nefza	1793	167
Local	1460	117

[1] National average for the last 15 years.
[2] Average of 3 sites during 1982–87.

adopting better agronomic practices such as planting at the appropriate time. Data indicated that a 69% and 117% increase over the present national average yield could be achieved for spring sown local cultivars of chickpea and lentil respectively. This increase could be improved even further by using better yielding cultivars such as Kassab for chickpea and Nsir and Nefza for lentil. The use of winter planting in chickpea could lead to a substantial yield increase ranging from 154 to 243%, but *Ascochyta* resistant varieties must be used; this is an essential condition.

Forage Legumes

The beneficial role of the forage legumes in any type of rotation has often been associated with their residual effect and its impact on the yield of the wheat grown after them. This impact varies with the legume species and it has been shown (Laboratoire des Cultures Fourrageres 1975) that the largest positive effect is given by clover (*Trifolium*), followed by sulla (*Hedysarum*) and lucerne (*Alfalfa*) (Table 9 trial 1 – 3). The vetch and oat

Table 9. Wheat grain yield (t/ha) following various forage legumes.

Previous crop	Grain yield	Straw yield
	Trial 1	
Vetch-oat mixture	3.36	1.33
Clover	4.00	1.38
Cereal	1.88	1.15
	Trial 2	
Vetch-oat mixture	1.13	
Lucerne (two years)	1.91	
Lucerne (three years)	2.13	
Lucerne (four years)	2.05	
Fallow	1.97	
Cereal	1.17	
	Trial 3	
Sulla (one year)	2.58	
Sulla (two years)	2.77	
Vetch-oat mixture	1.70	
Fallow	2.00	
Cereal	1.60	

Source: Seklani (1987).

mixture seems to have a rather detrimental effect on wheat yield when compared to a clean fallow.

Annual Medic

As indicated earlier, ley farming was introduced in the early seventies and since then an extensive research program has been conducted. Variety screening was undertaken and two varieties (*Medicago truncatula* cv Jemalong and *Medicago littoralis* cv Harbinger) were found to be the best adapted to Tunisian growing conditions. Dahmane (1987) mapped the rainfed northern part of Tunisia for *Medicago* suitability in relation to rainfall (Fig. 2) and Seklani (1981) did the same work in relation to other adaptation factors such as those indicated in Figure 3.

The effect of medic on wheat yield in comparison with fallow can be seen in a two-year rotation study conducted by the Office des Cereales on farmers' fields. The results clearly show a substantial yield increase as the

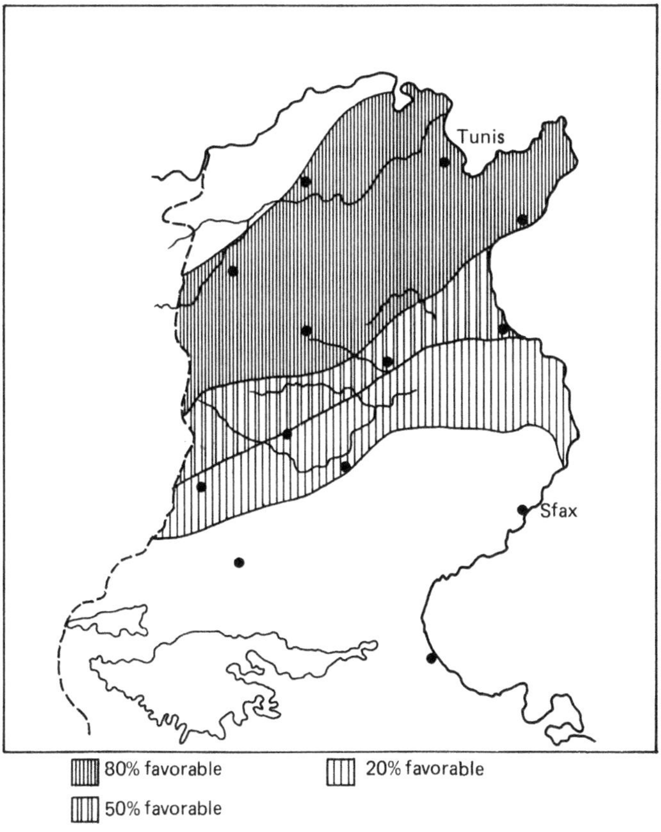

Fig. 2. Zone classification for growing *Medicago* in relation to yearly climatic conditions.

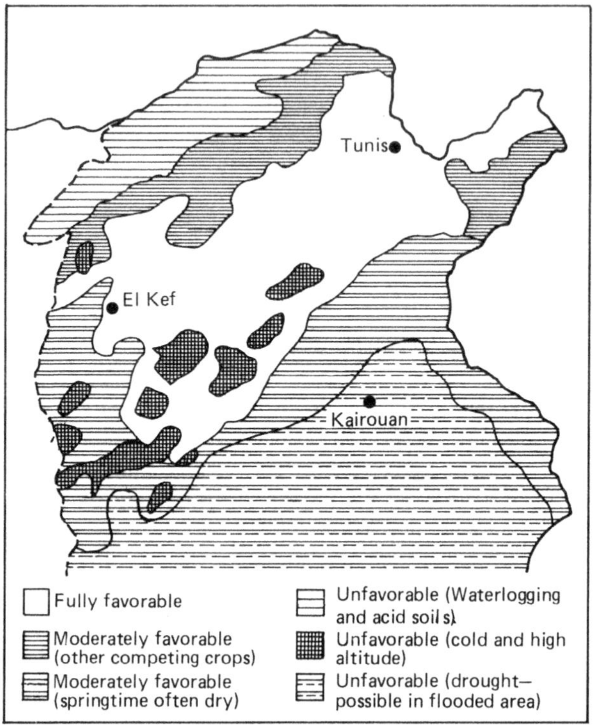

Fig. 3. Classification of zones in northern Tunisia according to suitability for *Medicago*.

result of medic, apparently due to improvement in soil nitrogen fertility (Table 10). The contribution to soil fertility by medic has been confirmed by a long term study (Ouertani and Dahmane 1987) where growing medic has resulted in a 12% increase in total soil nitrogen of the first 20 cm depth compared with a fallow treatment (Table 11). However, in another study conducted by Office des Cereals on farmers' fields the wheat grain yield after medic was exactly the same as after fallow. The explanation suggested

Table 10. Comparative effect of medic-ley and clean fallow on wheat grain yield (t/ha) in various locations with or without nitrogen fertilizer (kg/ha).

Crop rotation	1981/82[1] Nitrogen fertilizer			1984/85[2] Nitrogen fertilizer		
	0	33	66	0	33	66
Medic-wheat	1.672	1.860	1.526	2.746	2.064	3.248
Fallow-wheat	1.377	1.526	1.583	2.315	2.628	3.100

[1] Average of seven sites with rainfall 350–500 mm.
[2] Average of two sites with rainfall 400–500 mm.

Source: Office des Cereales–Technical division (1982 & 1985).

Table 11. Comparative total soil nitrogen (%) accumulated under medic and fallow cropping systems under semi-arid conditions of Tunisia, 1981–86.

Rotation	Soil depth (cm)		
	0–20	20–40	40–60
Fallow–wheat	0.107	0.083	0.073
Medic–wheat	0.120	0.088	0.082
Difference	0.013*	0.005*	0.009*

* Significant at 5% level.

Source: Ourtani and Dahmane (1987).

was that farmers do not usually apply phosphate fertilizer to their medic in rotation with cereals and therefore nitrogen fixation was not adequate. The effect of phosphate on nitrogen fixation by legumes has been demonstrated by Abbot and Robson (1984) and Dahmane (1987). Results of trials conducted in Tunisia (Gachet 1979; Khaldi et al. 1986) also show a similar positive response of legumes (medic and faba bean) to phosphate application. This response is reflected in the grain yield of wheat grown after the two legumes (Table 12).

Moisture profile studies under bare fallow and under medic were conducted by Ouertani and Dahmane (1987). Results of their studies indicated that soil moisture content at the end of the growing season (June) was higher under fallow than under medic. However, no difference was noticed in September, after the summer (Fig. 4).

The impact of the introduction of medic on agricultural production is illustrated by Table 13, where the average number of sheep was highly correlated with acreage of medics on 10 farms studied in northern Tunisia (Dahmane and Rhimi 1979). Similarly Table 14 shows lamb meat gain as the result of the introduction of medic pasture. This gain is also expressed in terms of wheat equivalence (t/ha). The two tables indicated clearly that the

Table 12. Comparative wheat yield (t/ha) following medic and faba bean with or without phosphate fertilizer at Goubeliat 1984–86.

Previous crop	Phosphate (P_2O_5 kg/ha)	Wheat yield[1]		Estimated gain in soil nitrogen, in kg of NH_4 NO_3 (33%)
		Trial 1	Trial 2	
Medic	0	0.64	0.72	60–90
Medic	90	0.84	1.54	
Faba bean	0	0.60		115
Faba bean	45	1.79		

[1] The wheat crop received 17 kg of nitrogen and 45 kg of P_2O_5.

Source: Khaldi et al (1986).

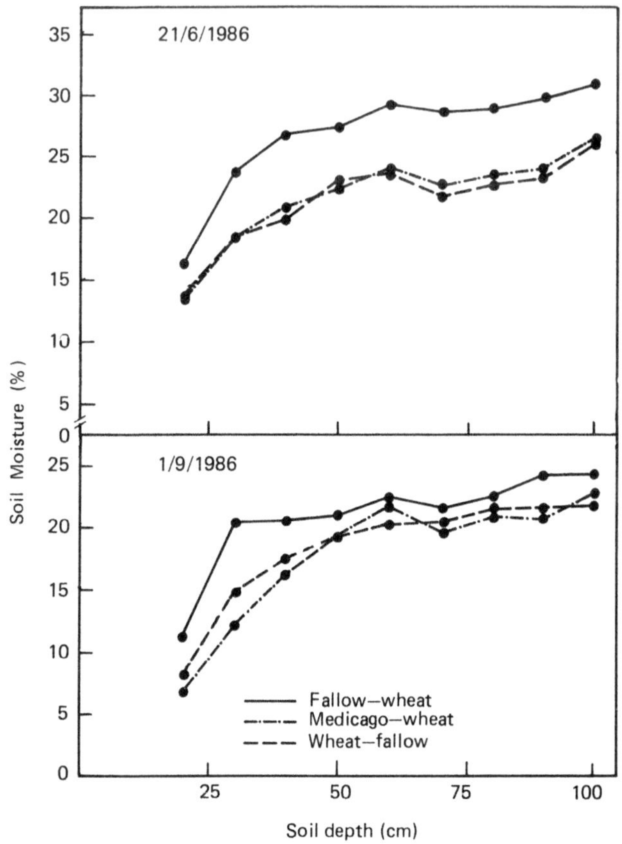

Fig. 4. Soil moisture profiles in June and September 1986.

Table 13. Relationship between area (ha) sown to medic and the number of sheep in 10 farms in northern Tunisia.

Year	Medic area	Number of sheep
1973	152	1660
1974	342	1812
1975	494	3952
1976	911	5410
1977	941	5948
1978	1055	6123
1979	1217	8260

Coefficient of correlation 0.80***

Source: Dahmane and Rhimi (1979).

Table 14. Impact of introduction of medic pasture on lamb meat production at two experimental sites (Oueslata and Bou Rebia).

Year	Meat gain (kg/ha)	Wheat equivalent gain (kg/ha)
1972/73	119–393	1.40–4.80
1974/75	143–319	1.70–3.90
1975/76	197–347	2.40–4.20

Source: Gachet (1979).

integration of livestock with cereal production by the use of the ley farming system is of paramount importance and remains the ultimate goal of the medic/cereal-based rotation.

Acknowledgement

The authors are grateful to Mr A. Haddad for the informative discussions on this subject, and for his input in updating the statistical data on acreage of annual medic.

References

Abbot, L.K. and Robson, A.D. 1984. The effect of VA mycorrhizae on plant growth. Pages 113–130 *in* VA Mycorrhizae (Powell, C.L. and Bagyaraj, D.J., eds). CRC Press, Boca Raton, Florida, USA.

Dahmane, A.B.K., Doolette, J.B., and Meddeb, A. 1972. Introduction des medicagos annuels dans les rotations cerealieres en Tunisie. Rapport annuel. Office des Cereales. Division Technique. Ministere de l'Agriculture, Tunis, Tunisia

Dahmane, A.B.K. and Rhimi, I. 1979. Le systeme d'assolement biennal medicago/ble. Son introduction et son evolution en Tunisie. Memoire de fin d'etudes du 2eme cycle. INAT, 43 Avenue Charles Nicolle, Tunis, Tunisia.

Dahmane, A.B.K. 1982. Principes fondamentaux des rotations medicago annuel/cereale dans les zones semi-arides. Pages 167–179 *in* Special report 668 AESCSD/OSU. Tunisia/USAID cereal breeding and production symposium. Tunis, Tunisia, April 12–16 1982.

Dahmane, A.B.K. 1987. Role of annual medicago species in the improvement of cereal and livestock production in Tunisia. A research review. INAT, Tunis, Tunisia.

Donald, C.M. and Williams, C.H. 1954. Fertility and productivity of a podsolic soil as influenced by subterranean clover (*Trifolium subterraneum* L.) and superphosphate. Australian Journal of Agricultural Research 5: 664–687.

Gachet, J.P. 1979. Rapport d'activities du Laboratoire des Cultures Fourrageres. INRAT, Tunis, Tunisia.

Haddad, A. 1982. Experimentation Medicago/Ble. Pages 179–189 *in* Special report 668 AESCSD/OSU. Tunisia/USAID cereal breeding and production symposium. Tunis, Tunisia. April 12–16, 1982.

Halila, H.M. 1986. Les legumineuses alimentaires. Pages 29–68 *in* L'Organisation de la recherche agronomique en Tunisie. Etude ISNAR/World Bank (ISNAR eds). The Hague, Holland.

Khaldi, R., Khaldi, G., Chakroun, M., Stilwell, T., Dahmane, A., Amara, H. 1986. Rapport technique sur les programmes et les resultats des recherches obtenus au cours de la campagne agricole 1985/86. Cooperation INRAT/INAT/ICARDA/CRDI (INRAT/INAT eds). Tunis, Tunisia.

Laboratoires des Legumineuses Alimentaires. 1986. Rapport d'activites 1985/86. INRAT-Ministere de l'Agriculture, Tunis, Tunisia.

Laboratoire des Cultures Fourrageres. 1975. Rapport d'activites 1972–75. INRAT-Ministere de l'Agriculture, Tunis, Tunisia.

Meeting the Challenge. Cereal Improvement in Tunisia. 1988. INRAT/ICARDA. ICARDA, Aleppo, Syria.

Ministere de l'Agriculture. 1986. VI plan (1982–1986) de developpement economique et social. Republique Tunisienne, Tunis, Tunisia.

Office des Cereales. 1988. Technical Division. Rapport annuel d'activites pour les annees 1971–87. Office des Cereales. Republique Tunisienne, Tunis, Tunisia.

Ouertani, A. and Dahmane, A. 1987. Etude comparative de l'influence a court terme des rotations jachere travaillee/ble et medic/ble sur l'humidite du sol et ses proprietes physico-chimiques. Memoire de specialisation. 3eme cycle. INAT. Ministere de l'Agriculture, Tunis, Tunisia.

Seklani, H. 1981. Place des luzernes annuelles dans une intensification fourragere. Document de Laboratoire des Cultures Fourrageres. INRAT, 2080 Ariana, Tunisia.

Discussion

Ibrahim:
Has there been any detailed economic and social study to explain why the ley farming system is not expanding in large areas, and to find out the attitude of farmers towards adoption of the system?

Halila:
No detailed study has been made.

Abd El Moneim:
Do you have grazing management trials to guarantee enough stored seeds for generation, and do you advise farmers on the right time for grazing and carrying capacity?

Halila:
A good deal of research was done on grazing and carrying capacity with suitable medics, and a fairly good technical package is available to farmers.

Papastylianou:
How good is re-establishment of medics (not reseeded) in the third year, after the cereal?

Halila:
Fairly good in the case of good management of the medics, which is not often the case.

Abd El Moneim:
Do you use Australian cultivars, or local germplasm from native habitats?

Halila:
Medic cultivars recommended to farmers are Australian, mainly Jemalong and Harbinger. Local germplasm collection and evaluation is now in progress.

Durutan:
Do you have any comment on the weed density of winter-sown chickpea or lentil? Do you think that winter planting will be increased, as expected, in the absence of an effective herbicide?

Halila:
In winter-sown food legumes weed infestation could be very severe and full crop loss could occur. I agree that if winter planting of chickpea is going to be pushed we must be able to provide the farmer with adequate weed control methods.

Tawil:
Did you select any lines suitable for winter planting from local chickpea? How susceptible is local chickpea to ascochyta blight?

Halila:
The local chickpea populations are highly susceptible to ascochyta blight and not suitable for winter planting. However, they show good yield potential under winter planting if they are protected against disease.

The Role of Legumes in the Farming Systems of Turkey

M. GULER

Field Crops Improvement Center, P.K. 226, Ulus-Ankara, Turkey

Abstract. The total area devoted to field crops in Turkey is approximately 24 million hectares. 5.8 million hectares of that is under fallow and 18.2 million hectares are occupied by various field crops. Food and fodder legumes are the second most important crop group, following cereals, occupying 11% of the field crop acreage, which is about 2 million hectares.

Approximately three-quarters of the overall legume acreage is occupied by lentil, chickpea and vetch. Turkey is the second country in the world in terms of lentil production and third in chickpea. It produces around 20% of world lentils and 5% of chickpeas. Exports in 1986 were more than 350 000 tons for both lentil and chickpea. Food legumes are cultivated at relatively lower altitudes, and fodder legumes are grown in higher and more mountainous areas of the country.

Drills and combines are used only in the south-eastern region where the climatic conditions and the size of the farms make this possible.

Research is being carried out according to the national projects on food legumes and fodder crops, and close linkages between research and extension help to transfer the research information to the farmers.

Introduction

The total area devoted to field crops in Turkey is approximately 24 million hectares, 5.8 million hectares is fallow and 18.2 million hectares occupied by various field crops. Cereals are the most important crops, covering 76% of the field crops acreage. Food and fodder legumes are the second most important crop group, together occupying 11% of the field crop acreage (Fig. 1) which is roughly 2 million hectares.

Approximately three-quarters of the overall legume acreage is occupied by lentil, chickpea and vetch (Fig. 2).

The total area planted to food legumes (lentil, chickpea, faba bean and pea) has increased gradually from 1970 to 1986 (Fig. 3). The sharp increase in the food legume acreage after 1981 is mainly due to the ongoing national project, 'The Utilization of Fallow Areas', which has been implemented since 1980. It is also possible to see a slight increase in the fodder legume area after 1981.

A.E. Osman et al. (eds.), The Role of Legumes in the Farming Systems of the Mediterranean Areas, 131–139.

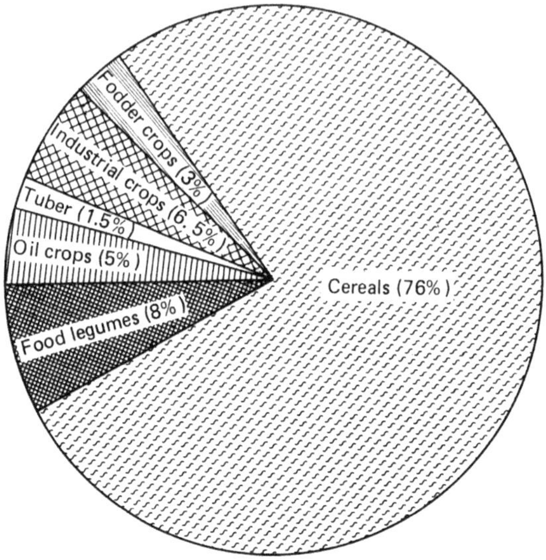

Fig. 1. Distribution of field crop acreage, 1986.

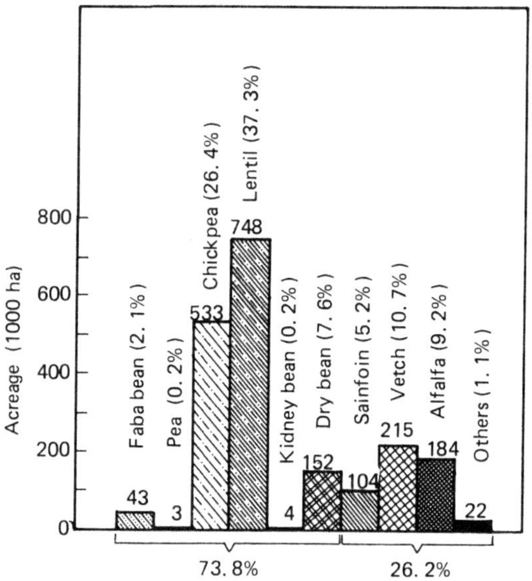

Fig. 2. Distribution of food and fodder legumes, 1986.

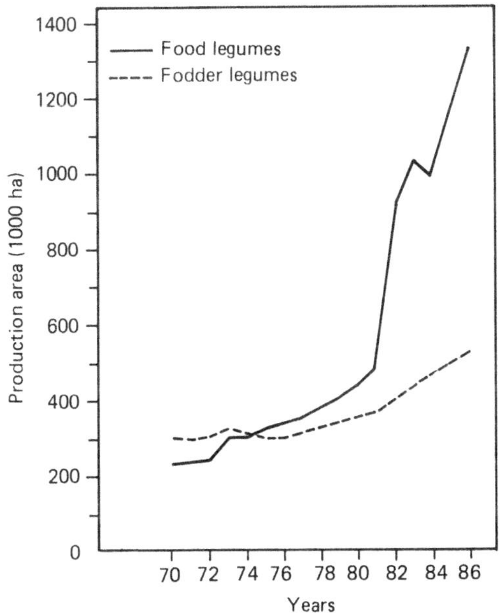

Fig. 3. Overall food and fodder legume acreage, 1970–86.

When each food legume crop is taken individually (Fig. 4), it can be seen that the main increases in area occurred in lentil and chickpea, whereas pea and faba bean acreage hardly changed.

When fodder legumes are taken into consideration (Fig. 5), a similar slight increase in area can be observed for vetch, alfalfa and sainfoin.

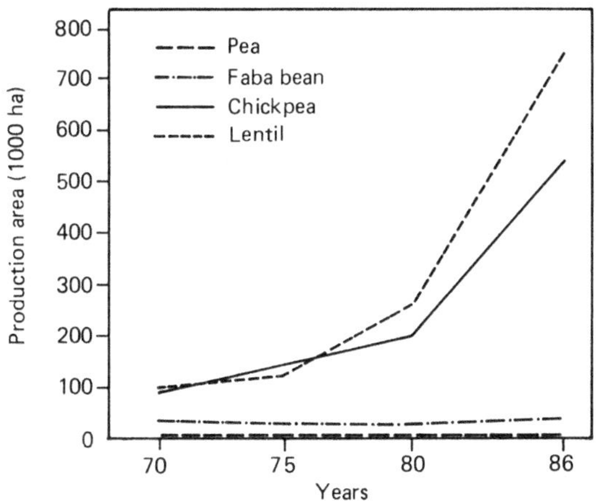

Fig. 4. Pea, faba bean, lentil and chickpea acreage, 1970–86.

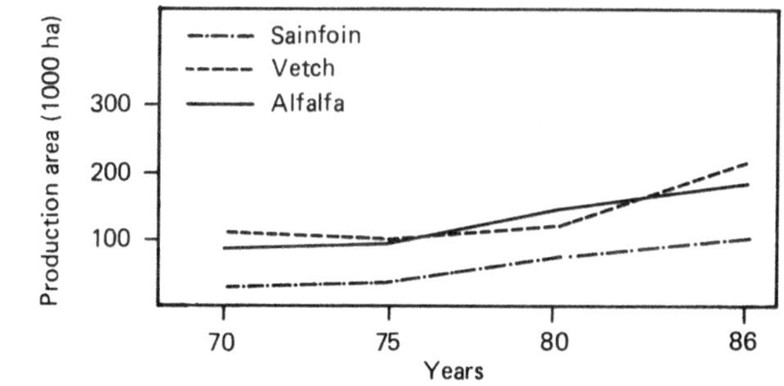

Fig. 5. Vetch, alfalfa and sainfoin acreage, 1970–86.

Turkey is the second largest producer of lentil after India, and third largest chickpea producer after India and Pakistan. It produces approximately 20% of the world lentils (850,000 t) and 5% of chickpea (630,000 t). Exports of both lentil and chickpea over the last 11 years reached 350,000 t, an increase of more than 300,000 t (Fig. 6).

Physical Environment

Climate

It is possible to divide the legume-producing areas of Turkey into five main regions according to the average temperatures (Table 1 and Fig. 7).

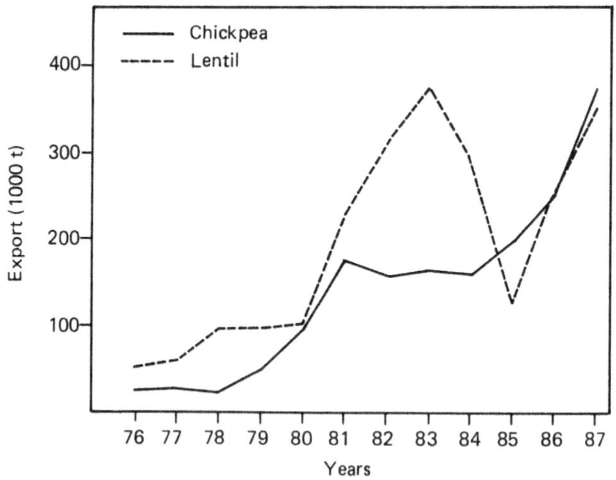

Fig. 6. Lentil and chickpea exports, 1976–87.

Table 1. Mean temperatures and elevation of the main legume-producing regions.

Region	Mean temperatures (°C) January	July	Elevation (m)
South coastal	7 – 10	27 – 28	1 – 100
South eastern	1.5 – 6	27 – 31	400 – 900
Marmara	1.5 – 6	22 – 24	1 – 1000
Central	0 – −4	19 – 23	800 – 1500
Eastern	−8 – −12	17 – 20	1300 – 1900

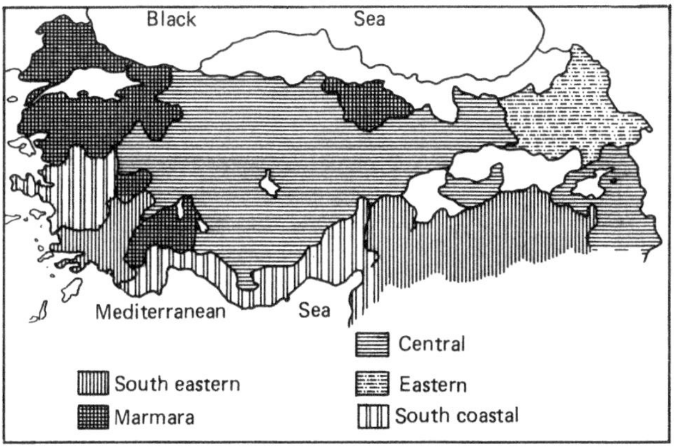

Fig. 7. The climate of legume-growing areas.

Mediterranean-type precipitation prevails all over the country; more than 65% of annual precipitation occurs in winter and spring. It is 300–500 mm in Central Anatolia and some parts of the south-eastern region, but higher in the rest of the country (Fig. 8).

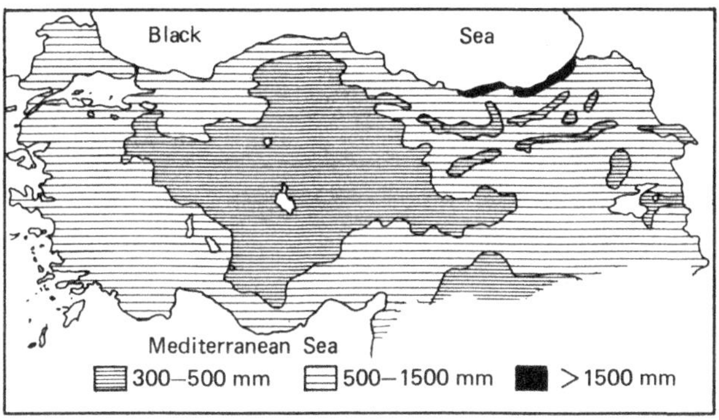

Fig. 8. Annual average precipitation patterns.

Present Role of Legumes in Farming Systems

When the food and fodder legume growing areas are examined region by region, it can be seen that food legumes are not being produced extensively in relatively colder and steeper areas, nor in industrially developed areas (Fig. 9). Fodder legumes are grown mostly in the higher mountainous areas of the country (Fig. 10).

Crop Sequence

Legumes are grown mainly in three environments: under dryland conditions in a cereal-legume rotation (most of the food legumes and vetch); in marginal areas (sainfoin); and under irrigated conditions (alfalfa).

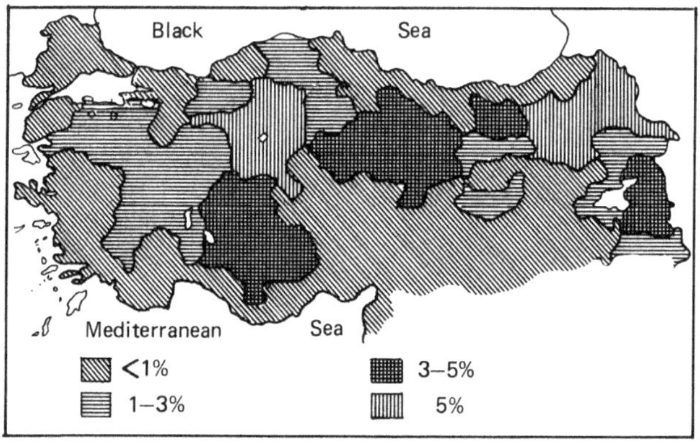

Fig. 9. Food legume production areas, 1986 (% of total area).

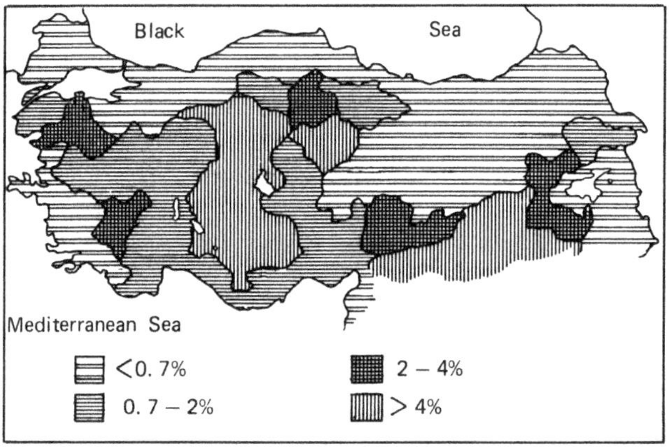

Fig. 10. Fodder legume production areas, 1986 (% of total area).

Mechanization

In Central and Eastern regions seeding is done in spring. In these regions mouldboard plowing after the broadcasting of seed on to the stubble is the main seeding method for all legumes.

In the South-eastern region, red cotyledonous lentil, which is the main legume crop of the area, is seeded in November. Farmers usually wait for the first rain and then prepare the seedbed by goble disc and seed by drill, or incorporate the seed through discing. Use of drills in this region has increased tremendously in recent years. Harvesting is being done by hand in the Central and Eastern regions where the sizes of the farms are relatively small. In the South-eastern region, the number of farmers using combine harvesters is increasing gradually. Combines in this region are being used on larger farms.

Constraints

It is possible to examine the constraints limiting productivity of the main legumes under two growing conditions according to the regions.

In the Central and Eastern regions, vetch and lentil are being seeded in spring, although research done in Central Anatolia proved that it is possible to double the yields by seeding in autumn with appropriate weed control. The main reason for not seeding in autumn is not the lack of suitable varieties: there are several cold-resistant varieties that can be used quite safely in Central Anatolia. It is that autumn seeding of lentil and vetch results in a serious weed problem in the spring. Use of herbicides for grassy weeds or volunteers is not a common practice. No herbicide is available for broadleaved weeds in legumes. In order to reduce the weed problem in the region, farmers usually broadcast the seed on stubble and then incorporate it by plowing as soon as the soil is ready in early spring. Plowing during seeding reduces weed density. In this system drilling is not preferred, since seedbed preparation is time-consuming and farmers do not want to risk seeding in the rainy season.

In the South-eastern region, the main legume is the red cotyledonous lentil. Seeding is done in late autumn. Soil is disced several times after rain, then the lentil is sown, mostly by drill. Weeds do not present a very big problem under this system. Relatively higher temperatures in winter enhance the continuous growth of weeds. The density of spring weeds is very low. In this case, discing several times after rain destroys the germinated weed seedlings.

Chickpea is mostly grown in Central and Western Transitional Anatolia. The main problem is late seeding in spring in order to escape from *Ascochyta* blight. If chickpea cultivars with reasonable cold tolerance and resistance to *Ascochyta* blight were developed, it would be possible to plant chickpea in autumn. It seems that there is promising genetic variation for winter-hardiness and tolerance to *Ascochyta* blight in the available material tested to date.

Farmers rarely use fertilizers on annual legumes. No insecticide is being applied to lentil in the field for *Sitona* sp. and *Bruchus* sp. Fumigation in storage is more common.

Research

The national food legume research project was put into operation in the late 1970s. A multidisciplinary approach in food legume research, a successful linkage between research and extension, and an improved credit program for the utilization of fallow areas resulted in a dramatic increase in food legume production.

A similar national project on fodder crop research has been in progress since the early 1980s. So far the project is still at the stage of establishment.

Extension of Technologies

We are trying to establish close linkages between research and extension organizations by conducting on-farm research and organizing field days, seminars and monthly workshops. Recent increases in legume area and production are the main result of these linkages.

Discussion

Dahmane:
Do you use any specific *Rhizobium* for chickpea? Is there any problem of nodulation of chickpea in winter compared to nodulation in spring?

Guler:
Rhizobium inoculation is not being used for chickpea since no problem has been detected in nodulation. Chickpea is only being grown as a spring crop in Turkey.

Solh:
Experience in Morocco in on-farm trials of winter chickpea indicated no difference in effective N_2-fixing nodulation between winter and spring in traditional production areas. In non-traditional production areas there is a need for inoculation with chickpea-specific *Rhizobia* strains.

Springborg:
Has there been a decrease in acreage planted to chickpea and lentil since the decline in export prices in 1985?

Guler:
I think the government will take necessary measures to keep the market, so no big changes are expected in food legume acreage in the short term.

Solh:
How critical is seed size in the local and export market for chickpea?

Guler:
Larger seeds are strongly preferred by both local consumers and export markets.

Jones:
Can you explain the reasons for implementing a fallow replacement policy? Is it just to increase the area of cropped land, or for better soil management? Can you comment on its effects on soil water management?

Guler:
The main idea was to get 2 crops in two years in the areas where there is enough precipitation, without decreasing the succeeding wheat yield, i.e. in areas with annual precipitation more than 350 mm, and/or shallow soils, and/or high spring and summer temperatures, where fallow efficiency is low. In higher rainfall areas (>350 mm) crop production does not depend on stored moisture.

Gintzburger:
Have annual legumes such as medics any potential for future use in Turkey?

Guler:
In order to increase livestock production I strongly believe that we have to integrate cereal production and livestock production. In keeping livestock in the system we can utilize the great potential of such annual legumes.

Gintzburger:
Do you use sainfoin in Turkish farming systems?

Guler:
Yes. We are encouraging its use in the areas where the slope is more than 8% which is mostly under cereals at present.

PART II

THE ROLE OF LEGUMES IN HUMAN AND ANIMAL NUTRITION

The Role of Food Legumes in the Diets of the Populations of Mediterranean Areas and Associated Nutritional Factors

P.N. BAHL

Division of Genetics, Indian Agricultural Research Institute, New Delhi - 110012, India

Abstract. Food legumes have an important role to play not only in increasing the quantity of food for Mediterranean populations but in improving the quality of their cereal-based diets. This paper highlights the different nutritional aspects, including the anti-nutritional factors, of food legumes, and attempts to summarize the available literature on the nutritional importance of faba bean, lentil, chickpea and peas for the people of the Mediterranean areas, particularly in the less-developed countries.

Introduction

Food legumes together with cereals account for as much as two-thirds of the total dietary intake of the vast populations of many developing countries. Food legumes, which contain two or three times as much protein as cereals, offer the most practical means of eradicating protein malnutrition in the cereal-based diet of the populations of Mediterranean areas. This implies that protein availability in these areas should be boosted by increasing the supply of food legumes rather than organizing costly feeding programs based on protein-rich foods of animal origin, or fortification of traditional foods with amino acids like lysine (Bahl et al. 1980). However, the low yield potential of the existing cultivars, together with instability of yield from year to year due to biotic and abiotic stresses, still remain major constraints for increasing availability of food legumes. Efforts should therefore be made to improve substantially the yield of these crops which, in turn, would result in higher availability of proteins per unit area.

Nutrient Composition

Among vegetable foods, legumes are nutritionally important because they are rich in proteins. The protein content of lentil, pea, faba bean and chickpea varies from 18% to 32% (Table 1).

Protein is mainly located in the cotyledons and embryonic axis of the grain. The cotyledons, embryonic axis and seed coat contain 27%, 48% and 5% protein, respectively (Bressani 1973). Actual protein content varies

A.E. Osman et al. (eds.), The Role of Legumes in the Farming Systems of the Mediterranean Areas, 143–149.

Table 1. Variation in the protein content in different food legumes.

Legume	Crude protein %
Lentil	24.8–32.1
Pea	21.0–32.3
Faba bean	19.3–27.8
Chickpea	18.0–28.1

Source: Kapoor et al. 1972; Gupta 1982.

depending upon variety, soil type, climate, location, management practices etc. The quantity of protein is also influenced by germination. Crude protein content of dry peas, lentil, faba beans and mung beans was found to increase three days after germination (Fordham et al. 1975).

Carbohydrate content of food legumes is high (>50%) and energy per unit weight provided by them is equal to that of cereals. Though different legumes differ in their carbohydrate content, starch is always the major constituent.

Food legumes have been reported to be a good source of minerals like calcium, phosphorus, iron, copper and molybdenum. Their total mineral matter ranges between 3.0 and 4.5%.

Legumes are also a rich source of vitamins. A large number of researchers have studied these vitamins, which include thiamine, riboflavin, nicotinic acid, ascorbic acid, carotene etc. These studies have been reviewed recently by Gupta (1988). Some of these vitamins are found in larger amounts in germinated seeds and sprouts than in dry seeds. For example, in sprouted peas the amount of ascorbic acid increased by 20 times, riboflavin by 2.5 to 4.5 times (Chen et al. 1975) and thiamine by six times (Fordham et al. 1975) compared to dry seeds. These observations indicate the ancient wisdom of the people of Mediterranean areas who often include sprouted legumes in their breakfast.

Protein Quality and Amino Acids

The quality of a protein depends on factors like composition of amino-acids, biological value and digestibility. Food legume proteins are rich in lysine and poor in sulphur-containing amino-acids and tryptophan. Methionine is reported to be the first limiting amino-acid in all the legumes. The relative proportion of essential amino-acids in legumes important in this part of the world is given in Table 2.

From the point of view of human dietary requirements, food legumes provide an amino-acid pattern that complements that of cereal and together they improve the overall protein quality of the diet. Several weaning foods, based on a judicious combination of cereals, legumes and oil seeds have been suggested to supply balanced diets in the developing countries to alleviate malnutrition.

Table 2. Essential amino-acid composition in food legumes (g/100g protein).

Amino Acid	Beans	Lentil	Chickpea	Peas
Lysine	6.8	5.1	6.3	8.9
Threonine	3.3	3.0	3.4	4.2
Valine	5.4	5.1	5.5	6.5
Leucine	8.9	5.5	8.2	9.5
Isoleucine	6.0	5.8	6.0	7.4
Methionine	1.0	0.6	1.2	1.3
Tryptophan	1.0	0.6	0.8	0.7
Phenylalanine	5.5	4.0	4.9	4.6
Arginine	9.2	7.0	6.9	13.4
Histidine	2.8	2.1	2.3	2.7

Source: Patwardhan and Ramachandran (1960); Banerjee (1960).

Nutritive Value of Legume-cereal Based Diet

Food legumes contain high amounts of lysine but are deficient in sulphurs amino-acids. In contrast to this, cereal grain proteins are low in lysine but have adequate amounts of sulphur-containing amino-acids. Evidently, in terms of pattern and profile of amino-acids, legume grain protein appears to be supplementary to cereal grain protein. Based on these considerations, it is important to know the role of food legumes in improving the nutritional quality of the cereal-based diets.

Protein quality has been found to be improved by the use of mixtures of cereals and legumes as compared to cereal grain alone (Table 3). In one such mixture consisting of maize and beans (*Phaseolus vulgaris*) with each contributing 50% of the proteins, the PER (Protein Efficiency Ratio) value increased by 122.2 per cent. In other mixtures of cereals and legumes increase in quality ranged from 16.4 to 64.7%. Based on these results, Bressani (1973) suggested that cereal-based diets would be of higher protein

Table 3. Protein value of optimum cereal:legume mixtures.

Distribution of protein in diet			Protein Efficiency Ratio	
Cereal	Legumes	Cereal:legume %	PER	Increase %
Rice	beans	100:0	2.25	
		80:20	2.62	16.4
Maize	beans	100:0	0.90	
		50:50	2.00	122.2
Maize	cowpea	100:0	1.22	
		50:50	1.84	50.8
Wheat	beans	100:0	1.05	
		90:10	1.73	64.7

Source: Bressani (1973).

quality if the legume-grain intake could be increased. Patwardhan and Ramachandran (1960) found that the nutritive value of dietary proteins was higher when pulses and cereals were mixed in the proportion of 3:7 than when either one of them is consumed alone.

Hallab et al. (1974) reported that the nutritive value of Arabic bread was significantly improved when it was supplemented with 10, 20, 30, 40 and 50% chickpea flour. Angel and Del Sotelo (1978) found that mixtures of chickpea and cereals in various proportions had higher nutritional value than when chickpea or cereal formed the sole source of proteins. Supplementation of traditional Central Asian bread ('Colio-non') with 10–15% chickpea was found to improve the nutritive value of the cereal-based diets (Makhmudov 1980).

An 80:20 blend of wheat and chickpea, when fortified with vitamins and minerals was found to be a good supplement for growing children (Daniel et al. 1969). Similarly, intake of protein foods made from blends of chickpea, groundnut, and soybean flours when fortified with minerals and vitamins were found to be beneficial in improving the nutritional status of school-going children (Daniel et al. 1969; Guttikar et al. 1965a and b). In short, increasing availability of food legumes through production gains and adding more to the diet would be cheaper than resorting to costly fortification of traditional foods with vitamins and minerals.

Legume-protein Digestibility and Related Anti-nutritional Factors

Digestibility of vegetable proteins is reported to be lower than animal proteins. In the case of food legumes it is known to be poor because of the presence of protein inhibitors which interfere in the process of protein utilization. Legume foods are often subjected to some degree of heat treatment before being eaten and digestibility has been found to improve considerably by different processing practices including cooking, although excessive heat reduces the nutritive value of proteins. Protein digestibility in cooked food legumes varies from 50–60% in faba beans to 80–93% in lentils (Table 4).

Table 4. Apparent protein digestibility (%).

Legumes	Digestibility
Lentil	80–93
Chickpea	76–90
Pea	71–94
Faba bean	50–60

Source: *a*. Nutritional standards and methods of evaluation for food legume breeders IDRC-TS7e, Ottawa, Canada, 1977; *b*. Nutritional improvement of food legumes by breeding. Protein Advisory Group of United Nations, New York, USA, 1973.

Anti-nutritional Factors

Since legumes are being considered as a supplement to cereal-based diets, it is essential to consider anti-nutritional and toxic substances which are usually found in legume seeds and are known to cause digestive disturbances. Many of these chemical substances like trypsin inhibitors, hemag glutinins, goitrogens, saponins, alkaloids, cyanogenetic glucosides etc. are thermolabile. Soaking and subsequent discarding of the water, and heat treatment can help in destroying the inhibitor activity of these toxic substances (Liener 1973).

Trypsin Inhibitors

The presence of trypsin inhibitors, which inhibit the proteolytic activity of certain enzymes, has been found in varying degrees in different food legumes. In faba beans, peas, lentil and chickpea, under identical assay conditions, trypsin inhibitor activity was found to be in decreasing order (Gallardo et al. 1974). Heat, moisture and mashing have been reported to be beneficial in destroying the inhibitor activity (Liener and Kakade 1980). Soni et al. (1978) reported that soaking in one per cent salt solution and sodium bicarbonate significantly reduced the trypsin inhibitor activity in pulses.

Phytohemagglutinins

Phytohemagglutinins or lectins have been identified in most of the food legumes (Liener 1979b). Lectins constitute two to ten per cent of the total protein in many legumes. These proteins have the property of agglutinating red blood cells. Hemagglutinins have been reported to be highly sensitive to heat treatment (Liener 1979a).

Polyphenols (Tannin)

Polyphenolic compounds are known to inhibit the activity of digestive enzymes in chickpea, more in cultivars with dark testa colour than in those with light testa colour (Singh 1984). In the case of chickpea, tannins were reduced by 60% when raw seeds were cooked and cooking water was discarded (Rao and Deosthale 1982). Tannins are also reported to be present in the seed coat of faba beans and they are known to inhibit proteolytic and other digestive enzymes (Griffiths 1979).

Favism

Favism is a disease characterized by hemolytic anaemia affecting certain individuals following consumption of raw or cooked faba beans. The presence of vicine and convicine occurring naturally as glycosides, and their hydrolytic products, divicine and isouramil in faba beans, act as the causative agents of favism (Beutler 1978).

Flatulence

Certain galactose-containing oligosaccharides, sucrose, stachyose and raffinose are responsible for the gas-producing factor in food legumes (Rackis 1975). However, the degree of flatulence varies with the species and varieties used (Calloway et al. 1971; Murphy 1972). Murphy (1972) further suggested genetic elimination of this undesirable trait by selection. This can help in the acceptance of food legumes in the diet of young children since in many segments of the population flatulence is associated with low digestibility.

Conclusions

There is an immediate need to improve the quality of the diets of the populations of Mediterranean areas while steadily increasing the quantity of available food. For alleviating protein malnutrition in humans, food legumes offer tremendous possibilities, because in terms of proteins they are complementary to cereals which constitute the major dietary intake among the nutritionally vulnerable groups. Careful processing, especially cooking, can improve the protein quality for human consumption. To improve digestibility efforts should be made to eliminate as far as possible anti-nutritional and flatulence factors by developing appropriate processing procedures. As a first step, high priority should be given to increasing the production and per capita availability of food legumes.

References

Angel, A.R. and Del Sotelo, A. 1978. Nutritional evaluation of cereal-legume mixtures using normal and genetically improved cereals and chickpea. Page 285 *in* Proceedings of the Fifth International Congress of Food Science and Technology, Kyoto, Japan, (Abstract), International Union of Food Science and Technology, John Wiley, Chicago, USA.

Bahl, P.N., Singh, S.P., Hayat Ram, Raju, D.E. and Jain, H.K. 1980. Breeding for improved plant architecture and high protein yields. Pages 297–307 *in* FAO/IAEA International Symposium on Seed Protein Improvement in Cereals and Legumes, Vienna, Austria.

Banerjee, S. 1960. Biological value and essential amino-acid composition of the proteins of some pulses. Pages 355–356 *in* Proceedings, Symposium on Proteins. Chemical Research Committee and Society of Biological Chemists, Mysore, India.

Beutler, M.D. 1978. Hemolytic anaemia in disorders of red cell metabolism (Winrobe, M.M., ed.) Plenum Press, New York, USA.

Bressani, R. 1973. Legumes in human diets and how they might be improved. Pages 15–42 *in* Proceedings of a Symposium sponsored by Protein Advisory Group of the United Nations (Milner, M., ed.), New York, USA.

Calloway, D.H., Hickey, C.A. and Murphy, E.L. 1971. Reduction of intestinal gas-forming properties of legumes by traditional and experimental food processing methods. Journal of Food Science 36: 251–255.

Chen, L.H., Wells, C.E. and Fordham, J.R. 1975. Germinated seeds for human consumption. Journal of Food Science 40: 1290–1294.

Daniel, V.A., Desai, B.L.M., Venkatrao, S., Swaminathan. M. and Parpia, H.A.B. 1969. Mutual and amino acid supplementation of proteins. IV. The nutritive value of proteins of blends of wheat and bengal gram fortified with limiting amino-acids. Journal of Nutrition and Dietetics India 6: 15.

Fordham, J.R., Wells, C.E. and Chen, L.H. 1975. Sprouting of seeds and nutrient composition of seeds and sprouts. Journal of Food Science 40: 552–556.

Gallardo, F., Araya, H., Pak, N. and Tagle, M.A. 1974. Toxic factors in Chilean legumes. II. Trypsin inhibitor activity. Archivos Latinoamericanos de Nutricion 24: 183–189.

Griffiths, D.W. 1979. The inhibition of digestive enzyme by extracts of field bean (*Vicia faba*). Journal of the Science of Food and Agriculture 30: 458–462.

Gupta, Y.P. 1982. Nutritive value of food legumes. Pages 287–327 *in* Chemistry and Biochemistry of Legumes (Arora, S.K., ed.). Oxford and IBH Publishing Co., New Delhi, India.

Gupta, Y.P. 1988. Nutritive value of pulses. Pages 561–601 *in* Pulse Crops (Baldev, B., Ramanujam, S. and Jain, H.K., eds.). Oxford and IBH Publishing Co., New Delhi, India.

Guttikar, M.N., Panemangalore, M., Narayanarao, M., Rajagopalan, R. and Swaminathan, M. 1965a. Studies on processed proteins in foods based on blends of groundnut, bengal gram, soybean, sesame flours and fortified with minerals and vitamins - I. Preparation, chemical composition and shelf life. Journal of Nutrition and Dietetics, India 2: 21–23.

Guttikar, M.N., Panemangalore, M., Narayanarao, M., Rajagopalan, R. and Swaminathan, M. 1965b. Studies on processed proteins in foods based on blends of groundnut, bengal gram, soybean, sesame flours and fortified with minerals and vitamins - II. Amino acid composition and nutritive value of the proteins. Journal of Nutrition and Dietetics, India 2: 24–27.

Hallab, A.H., Khatchadourian, H.A. and Jaler, I. 1974. The nutritive value and organoleptic properties of white Arabic bread supplemented with soybean and chickpea. Cereal Chemistry 51: 106–112.

Kappor, H.C., Srivastava, V.K. and Gupta, Y.P. 1972. Estimation of methionine in black gram (*Phaseolus mungo* Roxb.), green gram (*P. aureus* Roxb.) and soybean (*Glycine max* L.). The Indian Journal of Agricultural Science 42: 296–299.

Liener, I.E. 1973. Toxic factors associated with legume proteins. Journal of Nutrition and Dietetics, India 10: 303–322.

Liener, I.E. 1979a. Significance for humans of biological active factors in soybeans and other food legumes. Journal of the American Oil Chemists Society 56: 121–129.

Liener, I.E. 1979b. Protein inhibitors and lectins. Pages 97–122 *in* Biochemistry of Nutrition (Neuberger, A. and Jukes, P.H., eds.). University Park, Baltimore, USA.

Liener, I.E. and Kakade, M.L. 1980. Protease inhibitors. Pages 7–68 *in* Toxic Constituents of Plant Food Stuffs (Liener I.E., ed.). Academic Press, New York, USA.

Makhmudov, A. 1980. Increasing the nutritional value of national bread of the Central Asian Republics. Voprosy Pitaniya 6: 64–66.

Murphy, E.L. 1972. The possible elimination of legume flatulence by genetic selection. Pages 273–276 *in* Improvement of Food Legumes by Breeding. Proceedings of Symposium sponsored by Protein Advisory Group of the United Nations (Milner, M., ed.), New York, USA.

Patwardhan, V.N. and Ramachandran, M. 1960. Vegetable proteins in nutrition. Science and Culture 25: 401–407.

Rackis, J.J. 1975. Oligosaccharides in Food Legumes. Alphagalactosidase activity and the flatus problem. Pages 207–212 *in* Physiological Effects of Food Carbohydrates (Jeanes, A. and Hodge, J., eds). American Chemical Society, Washington DC, USA.

Rao, P.O. and Deosthale, Y.G. 1982. Tannin contents of pulses. Varietal differences and effect of cooking and germination. Journal of the Science of Food and Agriculture 33: 1013–1016.

Singh, U. 1984. Dietary fiber and its constituents in desi and kabuli chickpea (*Cicer arietinum* L.) cultivars. Nutrition Reports International 29: 745–753.

Soni, G.L., Singh, T.P. and Singh, R. 1978. Comparative studies on the effects of certain treatments on the antitryptic activity of the common Indian pulses. Journal of Nutrition and Dietetics, India 15: 341–345.

The Role of Food Legume Straw and Stubble in Feeding Livestock

B.S. CAPPER

Animal Feeds Section, Overseas Development Natural Resources Institute, 56/62 Gray's Inn Road, London, WC1X, United Kingdom; present address: International Livestock Center for Africa (ILCA), P.O. Box 5689, Addis Ababa, Ethiopia.

Abstract. The six million tonnes of food legume straws available in North Africa and West Asia can supply only 1% of the feed requirements of the ruminant livestock in these areas, compared with cereal straws which supply 30%. On a local or seasonal basis, however, food legume straw can be important, and an example is the use of lentil straw in the feeding of sheep during winter in Syria. Compared with barley and wheat straw, food legume straws contain more crude protein. Lentil and faba bean straws generally have higher metabolizable energy values than cereal straws but chickpea and pea straws often have lower values. Natural variation has been shown to exist in chickpea, lentil and faba bean straws. Straw from spring-sown chickpeas had a higher digestibility and crude protein content than that from winter-sown chickpeas but differences between chickpea varieties were not significant. Lentil varieties differed in the percentages of leaf, pod wall, branch and root material found in the straw. The digestibility of these plant fractions was in the order leaf>pod wall>branch>root. Variation in digestibility within fractions was considerable, particularly for pod walls. Biological methods, including nylon bag techniques, should be used in addition to conventional laboratory analyses to evaluate food legume straws. Where possible feeding experiments and animal production trials should be used in the evaluation process. Additional analyses for phenolic compounds, particularly phenolic acids esterified to cells walls, lignin and tannins may help to explain variation in straw quality and assist interpretation of animal-based investigations.

Introduction

The use of food legume straw and stubble in the feeding of livestock has received little attention from either plant or animal scientists. By contrast cereal straws have been intensively studied with attempts being made to improve their value by processing (Jackson 1977), supplementation (Preston and Leng 1984) and identifying cereal varieties with superior straw (Tuah et al. 1986). This paper examines the availability of food legume straws in the

A.E. Osman et al. (eds.), The Role of Legumes in the Farming Systems of the Mediterranean Areas, 151–162.

Table 1. Estimated availability of food legume straws[1] (1000 metric tonnes) and ruminant livestock[2] populations (million head) in some countries in the ICARDA region.

Country	Food Legume Straws				Livestock				Level in diets[3] (%)
	Chickpea	Lentil	Faba bean	Pea	Buffalo	Cattle	Sheep	Goats	
Morocco	71	140	335	83	–	2.6	12.1	4.7	3.59
Algeria	29	5	58	6	–	1.6	14.8	3.1	0.67
Tunisia	45	3	48	–	–	0.6	5.3	1.1	1.81
Libya	–	–	15	3	–	0.2	0.9	–	1.56
Egypt	26	40	517	8	2.6	2.8	2.6	2.7	2.74
Sudan	3	–	67	–	–	22.4	15.6	–	0.08
Cyprus	2	–	5	–	–	0.1	0.5	0.4	0.88
Turkey	900	2162	133	8	0.5	16.2	40.5	13.1	3.76
Syria	42	166	23	2	–	0.8	12.5	1.1	2.43
Lebanon	2	11	–	–	–	0.6	0.1	0.5	0.54
Jordan	3	11	2	–	–	0.1	1.1	0.5	1.39
Saudi Arabia	–	–	–	–	–	0.5	3.8	2.4	0.00
Yemen	–	–	–	–	–	0.9	1.9	2.3	0.00
Iraq	20	16	15	–	0.2	1.6	8.8	2.4	0.43
Iran	83	90	–	75	0.2	8.4	34.5	13.6	0.46
Afghanistan	–	–	–	–	–	3.8	20.0	3.0	0.00
Pakistan	867	69	–	105	13.4	16.7	25.8	30.8	0.78
Totals	2093	2713	1218	290	16.9	79.9	200.8	81.7	1.32

[1] Based on FAO (1987) and author estimates of harvest indices (chickpea, 0.40; lentil, 0.275; faba bean, 0.375; pea, 0.40).
[2] Populations less than 0.05 million not included.
[3] Estimated maximum overall inclusion level in ruminant diets.

ICARDA mandate region and their potential contribution to ruminant diets. The existing information on composition and feeding value is reviewed. Techniques which could be used to evaluate food legume straws and determine effects of variety or environment on quality are described.

Availability of Food Legume Straws

Faba bean straw is found mainly in North Africa, particularly in Morocco and the Nile Valley (Table 1). Chickpea and lentil straws are significant feed resources in Turkey and Syria. Chickpea straw is also found in large quantities in Pakistan but pea straw is available in substantial quantities only in Morocco, Iran and Pakistan. Food legume straws are not important overall in the diets of ruminant livestock in the ICARDA region. The maximum contribution to dry matter consumption is just over 1%. This contrasts with the potential overall contribution of cereal straws of about 30%. Countries where food legume straws are potentially important in relation to feed requirements are Morocco, Egypt, Turkey and Syria but the extent to which food legume straws are used for feeding livestock varies between countries. In Egypt much of the faba bean straw is used as fuel whereas in Syria lentil straw is conserved as an important feed for sheep (Nordblom and Halimeh 1982; Erskine 1983). Whilst food legume straws supply only a small part of the feed requirements of ruminant livestock, their importance may be considerable in certain agro-climatic zones within a country, and they may also be significant seasonally.

Role in Feeding Livestock

An example of the use of food legume straw in feeding livestock is in the feeding of sheep in winter in Syria. The extent to which legume straw, mostly lentil straw, is used varies according to agro-climatic zone (Table 2) with more being utilized in wetter zones than in some of the drier zones. On

Table 2. Use of legume straw in the winter feeding of sheep in Northern Syria (ICARDA, unpublished data, 1982).

Feed	Metabolizable energy intake (%) Zone (rainfall in mm)					
	>600	>250	250	>200	<200	Average
Cereal grains	51.0	42.7	64.4	38.1	47.1	48.7
Other grains	1.0	2.6	0.1	0.0	0.8	0.9
Cereal straw	3.2	18.3	22.7	39.1	22.8	21.2
Legume straw	32.0	17.0	3.1	0.0	10.5	12.5
By-products[1]	12.8	19.4	9.7	22.8	18.8	16.7

[1] Mainly cottonseed cake and wheat bran.

average 12.5% of the metabolizable energy supply comes from this source but in terms of dry matter consumed the proportion of legume straw in the diet will be around 15–20%.

Analysis of Food Legume Straws

Typical laboratory analyses for food legume straws are shown in Table 3 together with analyses of other feeds frequently used in Mediterranean areas. The crude protein content of food legumes straws is higher than cereal straws but less than that of hays, feed grains and many by-products. The phosphorus content, in common with cereal straws and hays, is low and would not meet animal requirements if provided as the sole feed. By contrast feed grains and by-products contain higher levels of phosphorus and have an important role in providing this element in addition to raising dietary crude protein levels so as to support milk, meat and animal fibre production.

Feeding Value of Food Legume Straws

The digestibility of lentil and faba bean straws measured during *in vivo* trials is generally higher than chickpea and pea straws which are often similar to cereal straws (Table 4). Allden and Geytenbeek (1980a) found that faba

Table 3. Laboratory analyses of feed legume straws in comparison with cereal straws, hays and concentrate feeds.

Feed	Analysis (%)					
	Crude protein	Crude fibre	Ether extract	Ash	Calcium	Phosphorus
Chickpea straw[1]	6.0	44.4	0.5	13.3	0.34	0.12
Lentil straw[2]	5.8	37.1	2.4	9.0	1.65	0.07
Faba bean straw[2]	5.7	33.7	0.8	14.1	na	na
Pea straw[2]	8.4	39.5	0.9	8.5	na	0.11
Barley straw[2]	4.2	35.6	2.1	14.6	0.49	0.16
Wheat straw[2]	3.8	38.9	2.0	12.6	0.47	0.34
Alfalfa hay[2]	16.7	35.0	1.9	9.0	2.08	0.22
Oat hay[2]	7.9	37.6	2.6	8.0	0.39	0.22
Vetch/barley hay[2]	11.2	27.4	4.0	10.2	na	na
Barley grain[3]	10.0	5.0	1.6	2.9	0.11	0.44
Wheat bran[3]	16.7	8.3	3.9	7.2	0.11	1.00
Cottonseed cake[3]	26.7	24.4	5.6	5.1	0.22	1.22
Soya bean meal[3]	50.0	6.1	1.1	6.1	0.33	0.67

[1] Gohl (1981).
[2] Arab and Middle East Tables of Feed Composition (1979).
[3] Parr et al. (1988).
na = no data available from reference used.

Table 4. Digestibility, metabolizable energy values and estimated levels of voluntary consumption in mature ruminants for food legume straws, cereal straws, hays and concentrate feeds.

Feed	Indicator of feed value		
	Digestibility[1]*	Metabolizable[1] energy value**	Voluntary consumption[2]***
Chickpea straw[3]	39.0	5.9	35
Lentil straw[4]	51.9	7.9	55
Faba bean straw	50.0	7.4	55
Pea straw	43.0	6.5	40
Barley straw[5]	45.0	6.8	40
Wheat straw	39.0	5.9	30
Alfalfa hay	51.0	7.7	60
Oat hay	50.0	7.8	55
Vetch/barley hay[6]	57.2	8.6	68
Barley grain	86.0	13.1	90–100
Wheat bran	61.0	10.1	50
Cottonseed cake	51.0	8.7	40
Soya bean meal	79.0	12.3	90–100

* g digestible organic matter. $100g^{-1}$ dry matter.
** megajoules. kg^{-1} dry matter.
*** g dry matter. kg liveweight $^{0.75}$. day^{-1}.

[1] Based on MAFF (1975) except where indicated.
[2] Author estimates.
[3] Gohl (1981).
[4] ICARDA (unpublished information, 1984).
[5] Capper et al. (1986).
[6] Rihawi et al. (1987).

bean straw has a dry matter digestibility (DMD) in sheep of 56.9 compared with 44.4 for barley straw. In a subsequent experiment (Allden and Geytenbeek 1980b) DMD values for faba bean and chickpea straws were 60.4 and 49.2 respectively. It appears that lentil and faba bean straws can be comparable in metabolizable energy value to some samples of alfalfa or oat hay. Hadjipanayiotou et al. (1985) found that both faba bean and chickpea straws were superior to the barley straws they evaluated. Thorlacius et al. (1979) found that faba bean straw intake and digestibility were superior to that of wheat straw and not significantly different from that of medium quality alfalfa-brome hay. Voluntary intakes of faba bean and chickpea straws were significantly greater than that of ryegrass hay (Allden and Geytenbeek 1980b).

In relation to the energy requirements of a ewe weighing 45 kg, feeds with digestibilities of 40, 50 and 60 would provide maintenance, maintenance plus 0.5 kg of milk, and maintenance plus 1.0 kg milk respectively. Lentil and faba bean straws could, if fed *ad libitum*, meet the requirements of ewes in late lactation. Similarly lambs of 25–30 kg in weight would grow at around 50 g per day if provided with these straws. Some food legume straws can be used potentially as sole feeds but normally they are fed at a relatively low level in the diet (10–15%) in combination with cereal straws, grains and by-products.

Variability in Food Legume Straw Quality

Chickpea

The winter sowing of chickpeas has been found to give higher seed yields compared with spring sowing. Table 5 gives laboratory analyses for chickpea straw of 16 varieties grown at the Tel Hadya site of ICARDA. Straw from spring sown chickpeas was significantly higher in digestibility and crude protein but lower in fibre than that from winter sown chickpeas. The effects of variety were not significant for any of the analyses carried out.

Lentil

Table 6 gives data on the composition of straw from 11 varieties of lentils indicating that particular varieties can have straw which is superior to that of other varieties. Some of the differences in straw digestibility could be caused by variation in the percentages of leaves, pod walls, branches and roots (Table 7). Entry number 3 had the greatest amount of branch material and the lowest straw digestibility. Entry number 2 had the lowest amount of branch material and the highest straw digestibility. Table 8 shows that the

Table 5. Analyses (%) of straw from winter and spring sown chickpeas (Rihawi and Singh, unpublished data, 1984)

Season	Digestibility[1]		Crude protein		Neutral detergent fibre		Ash	
	mean	se	mean	se	mean	se	mean	se
Winter	40.4	0.51	3.8	0.08	60.9	0.55	10.6	0.24
Spring	42.5	0.35	4.7	0.15	57.2	0.25	10.8	0.19
Significance[2]								
Variety	NS		NS		NS		NS	
Planting time	**		***		***		NS	

[1] g digestible organic matter 100 g^{-1} dry matter (Tilley and Terry 1963).
[2] NS, not significant; **, P > 0.01; ***, P > 0.001.

Table 6. Analysis (%) of straw from lentils (n = 11) and faba beans (n = 24)

Straw	Digestibility[1]			Crude protein			Neutral detergent fibre		
	mean	se	range	mean	se	range	mean	se	range
Lentils[2]	60.0	1.2	57.0–64.0	6.7	0.28	5.9–7.5	60.0	0.8	58.0–64.0
Faba bean[3]	44.6	0.6	39.9–51.9	5.0	0.13	4.0–5.6	65.7	0.8	56.4–72.7

[1] g digestible organic matter 100g^{-1} dry matter (Tilley and Terry 1963).
[2] Rihawi and Erskine, unpublished data, 1982.
[3] Rihawi and Robertson, unpublished data, 1983.

Table 7. Variation in the percentages of plant fractions in six lentil varieties and straw digestibility[1]

Entry number	Plant fraction				Straw[2] digestibility
	Leaves	Pod walls	Branches	Roots	
1	35.3	23.2	38.2	3.3	52.3
2	39.2	25.0	30.4	5.4	52.4
3	33.5	21.5	38.5	6.5	43.4
4	39.8	23.2	31.8	5.2	51.0
5	39.2	21.6	34.8	4.4	51.2
6	41.5	22.7	31.8	4.0	49.9
Mean	38.1	22.9	34.3	4.8	49.9
se	1.2	0.5	1.4	0.5	1.5

[1] Erskine, unpublished data, 1985.
[2] Pepsin/cellulase dry matter digestibility (Goto and Minson 1977).

Table 8. Mean analysis (%) of leaves, pod walls, branches and roots in six lentil varieties.

Fraction	Digestibility[1]			Crude protein			Neutral detergent fibre		
	mean	se	range	mean	se	range	mean	se	range
Leaves	66.2	1.1	62.5–70.4	8.6	0.3	7.9–10.0	29.2	0.8	26.1–31.0
Pod walls	48.4	2.2	42.7–57.2	6.3	0.2	5.9- 7.1	49.8	0.9	47.5–53.9
Branches	39.8	1.9	33.5–45.3	3.6	0.1	3.2- 3.9	60.2	1.8	55.1–67.7
Roots	25.9	1.6	20.1–30.5	6.2	0.2	5.8- 7.0	71.4	1.1	67.6–75.3

[1] Pepsin/cellulase dry matter digestibility (Goto and Minson 1977).

digestibility of fractions from highest to lowest was leaves, pod walls, branches and roots, thus entries with more leaf and pod wall material might have better straw digestibility than entries with more branch and root material. However, there was considerable variation in digestibility within particular plant fractions. The range in digestibility was greatest for pod walls. The causes of variation within plant fractions require further research.

Faba Bean

The data in Table 6 show that differences occur for straw digestibility in faba beans according to variety.

Evaluation of Food Legume Straws

Laboratory and Biological Techniques

Food legume straws can be evaluated by laboratory analysis, 'biological techniques', animal feeding experiments and production trials. Further laboratory analysis for phenolic compounds and other inhibitory substances may also be necessary. Conventional laboratory analysis can only provide a preliminary indication of feeding value. Crude fibre for instance does not include lignin which is virtually indigestible. Biological methods using cellulase (Goto and Minson 1977), rumen liquor (Tilley and Terry 1963) and nylon bags (Mehrez and Orskov 1977) may provide a better guide to differences between straws. An advantage of the nylon bag method is that the rate as well as the potential extent of straw digestion can be studied. Disadvantages of the nylon bag method are that material made soluble may not be digestible and indigestible material may pass through pores of the nylon bag during suspension in the rumen. Van Soest (unpublished method) has proposed use of a true digestibility system employing incubation apparatus similar to that used in the rumen liquor method but incorporating indicators to ensure that conditions are continuously anaerobic. Rates of digestion can be determined by using a series of tubes incubated for different times. Often this system is used to determine digestion of neutral detergent fibre.

Animal Feeding Experiments and Production Trials

Very few animal feeding experiments have been conducted with food legume straw. There is a need to determine the differences between straws from contrasting varieties in terms of voluntary intake and digestibility so that varieties with superior straw can be selected in breeding programs. This is particularly important for lentils where the financial value of the straw determines the economic viability of the crop in some instances (Nordblom and Halimeh 1982). There is also a need to determine the effects of supplements such as feed grains and by-products on food legume straw feeding value. Ultimately the use of food legume straw from different varieties in feeding pregnant, lactating and growing livestock could be studied.

Phenolic Compounds

Two types of phenolic compounds may affect the feeding value of food legume straws; those concerned with structural strengthening, lignin for example, and those involved in chemical defence and/or stress reactions (a range of polyphenolics including what are commonly called tannins). Lignin is composed of many phenylpropanoid units. Depending upon the level of substitution of methoxyl groups into the benzene ring the lignin polymer will contain varying amounts of coumaric acid, ferulic acid and diderulic acid. Ovejero Martinez (1967) has shown that the digestible energy content of food legume straws is negatively related to lignin content. A convenient way of estimating lignin is the permanganate method of Goering and Van Soest (1970). Unpolymerised phenolic acids, presumably lignin precursors, are esterified to cell wall carbohydrates and may also have an effect on digestibility (Hartley and Jones 1977).

Polyphenolics including tannins are present in many forage legumes. They may be important in determining food legume straw quality but there have been no investigations into this. Traditionally polyphenolics have been divided into condensed and hydrolysable tannins. Condensed tannins are formed by condensation of flavan-3-ols such as catechin or epicatechin and are the type normally present in forage legumes. The hydrolysable tannins contain a carbohydrate core to which phenolic carboxylic acids are bound by ester linkages. Most colorimetric methods of determining tannins suffer from non-specificity but a method which is acceptable for condensed tannins is that of Sarkar and Howarth (1976). A method for determining total phenolic content rapidly which holds promise is quantitative precipitation with ytterbium acetate (Reed, Horvath, Allen and Van Soest 1985). Separation of phenolics can be carried out with high performance liquid chromatography (for example Mueller-Harvey, Reed and Hartley 1987).

Conclusions

Food legume straws are quantitatively unimportant compared with cereal straws in the ICARDA region. However some may be important locally or seasonally and they can often support levels of animal production above maintenance without additional energy supplementation, provided protein and mineral requirements are met. Usually food legume straws are fed in combination with other feeds and form only a small proportion of the overall diet. Further work is required on the causes of varietal and environmental differences in straw quality. Application of biological methods, such as the nylon bag technique, animal feeding trials and investigations into the role of phenolic compounds in determining feeding value would help to enhance the use of food legume straws in feeding systems.

Acknowledgements

The collaboration of members of ICARDA's Food Legume Improvement Program, particularly Dr Erskine, and Mr Rihawi and staff in conducting analyses is gratefully acknowledged.

References

Allden, W.G. and Geytenbeek, P.E. 1980a. Assessment of field bean stubbles and supplements for grazing cattle and sheep. Proceedings of the Australian Society of Animal Production 13: 281–284.

Allden, W.G. and Geytenbeek, P.E. 1980b. Evaluation of nine species of grain legumes for grazing sheep. Proceedings of the Australian Society of Animal Production 13: 249–252.

Arab and Middle East Tables of Feed Composition. 1979. International Feedstuffs Institute, Utah, USA, and the Arab Centre for Studies of Arid Zones and Dry Lands, Damascus, Syria.

Capper, B.S., Thomson, E.F., Rihawi, S., Termanini, A. and MaCrae, R. 1986. The feeding value of straw from different genotypes of barley when given to Awassi sheep. Animal Production 42: 337–342.

Erskine, W. 1983. Relationship between the yield of seed and straw in lentil. Field Crops Research 7: 115–121.

FAO (Food and Agriculture Organization). 1987. FAO Production Yearbook 1986. FAO Basic Data Unit, Statistics Division, Rome, Italy.

Goering, H.K. and Van Soest, P.J. 1970. Forage fiber analyses (apparatus, reagents, procedures and some applications) Agriculture Handbook U.S. Department of Agriculture No. 379, Government Printing Office, Washington DC, USA.

Gohl, B. 1981. Tropical Feeds. FAO Animal Production and Health Series, No. 12. FAO, Rome, Italy.

Goto, I. and Minson, D.J. 1977. Prediction of the dry matter digestibility of tropical grasses using a pepsin cellulase assay. Animal Feed Science and Technology 2: 247–253.

Hadjipanayiotou, M., Economides, S. and Koumas, A. 1985. Chemical composition, digestibility and energy content of leguminous grains and straws grown in a Mediterranean region. Annales de Zootechnie 34: 23–30.

Hartley, R.D. and Jones, E.C. 1977. Phenolic components and degradability of cell walls of grass and legume species. Phytochemistry 16: 1531–1534.

Jackson, M.G. 1977. The alkali treatment of straws. Animal Feed Science and Technology 2: 105–130.

Mehrez, A.Z. and Orskow, E.R. 1977. A study of the artificial fibre bag technique for determining the digestibility of feeds in the rumen. Journal of Agricultural Science, Cambridge, UK. 88: 645–650.

Ministry of Agriculture, Fisheries and Food. 1975. Energy allowances and feeding systems for ruminants. Technical Bulletin 33. Her Majesty's Stationery Office, London, UK.

Mueller-Harvey, I., Reed, J.D. and Hartley, R.D. 1987. Characterization of the phenolic compounds, including flavanoids and tannins, of ten Ethiopian browse species by high performance liquid chromatography. Journal of the Science of Food and Agriculture 39: 1–4.

Nordblom, T. and Halimeh, H. 1982. Lentil residues make a difference. Lens 9: 8–10.

Ovejero Martinez, F.J. 1967. Energia digestible y metabolisable de las pajas de leguminosas para los ovidos. [Digestible and metabolizable energy of legume straws for sheep.] Anales de la Facultad de Veterinaria, Leon 13: 307–354. Nutrition Abstracts and Reviews 40: 252.

Parr, W.H., Capper, B.S., Cox, D.R.S., Jewers, K., Marter, A.D., Nichols, W., Silvey, D.R. and Wood, J.F. 1988. The small-scale manufacture of compound animal feed. Overseas Development Natural Resources Institute, Bulletin No. 9. ODNRI, London, UK.

Preston, T.R. and Leng, R.A. 1984. Supplementation of diets based on fibrous residues and by-products. Pages 373–413 *In* Straw and Other Fibrous By-products as Feeds (Sundstol, F. and Owens, E. eds). Elsevier, Amsterdam, The Netherlands.

Reed, J.D., Horvath, P.J., Allen, M.S. and Van Soest, P.J. 1985. Gravimetric determination of soluble phenolics including tannins from leaves by precipitation with trivalent ytterbium. Journal of the Science of Food and Agriculture 36: 255–261.

Rihawi, S., Capper, B.S., Osman, A.E. and Thomson, E.F. 1987. Effects of crop maturity, weather conditions and cutting height on yield, harvesting losses and nutritive value of cereal-legume mixtures grown for hay production. Experimental Agriculture 23: 451–459.

Sarkar, S.K. and Howarth, R.E. 1976. Specificity of the vanillin test for flavanols. Journal of Agricultural and Food Chemistry 24: 317–320.

Thorlacius, S.O., Coxworth, E. and Thompson, D. 1979. Intake and digestibility of faba bean crop residue by sheep. Canadian Journal of Animal Science 59: 459–462.

Tilley, J.M.A. and Terry, R.A. 1963. A two-stage technique for the *in vitro* digestion of forage crops. Journal of the British Grassland Society 18: 104–111.

Tuah, A.K., Lufadeju, E., Orskov, E.R. and Blackett, G.A. 1986. Rumen degradation of straw. 1. Untreated and ammonia-treated barley, oat and wheat straw varieties and triticale straw. Animal Production 43: 261–269.

Discussion

Ibrahim:
Food legume straw is so important in Syria and Turkey that it has to be taken into account when designing harvesting machinery. Is this the case in other countries? What is the importance of food legume straw, especially lentil, in North African countries?

Capper:
Food legume straws can only provide about 1% of the feed requirements of livestock in the Mediterranean areas, but they are important in certain areas, for example, lentil straw in Syria and Turkey, faba bean straw in Egypt. In general, it appears that very little research has been carried out on the feeding value of food legume straws or at least very little work has been published.

Bounejmate:
In Morocco it is very important but very expensive. Farmers value it highly.

Halila:
It is not used at all in Tunisia; I don't know why. There has not been enough research. It may be used as fuel in winter.

Jones:
If it isn't used it suggests that in the place where it is grown there aren't any hungry animals, or there is no market for sending it to the other end of the country. In Syria and Turkey animals are in the same areas as where crops are grown.

Akkada:
There are many hungry animals. An AOAD/ACSAD study of 21 Arab countries showed that self-sufficiency in feed resources is only 60%–65%. There is obviously a need to look into all crop residues and by-products.

Kamel:
Is it correct to assume that digestibility and crude protein are environment-specific, and what applies to rainfed conditions might not in irrigation systems?

Capper:
Cereals straw grown under rainfed conditions is generally more digestible than that from irrigated areas. Despite clear environmental effects it has been found that genotype effects are usually higher than genotype x environment interactions. The situation appears to be different for food legume straws where environmental effects and G x E interactions are both more important than genotype effects. Whereas it may be possible to select varieties of cereals with superior straw quality, it may not be possible in food legumes where environmental effects appear to be paramount.

Solh:
To what extent are the levels of palatability and digestibility correlated?

Capper:
It is necessary to distinguish two separate effects. The first is the effect of particle size and rate of digestion on rumen fill and hence voluntary intake. The second is differences in organoleptic quality between straws, caused by physical factors and smell. For the first effect there will be some correlation but this may not occur with the second effect.

Solh:
Most farmers growing large-seeded faba bean in West Asia and North Africa harvest the green pods two or three times throughout the season then the harvested fields are leased for grazing. How would processing of the mature faba bean plants improve their palatability?

Capper:
Chopping or grinding them would increase consumption but would have only a small effect on digestibility. Investigations on the stage of maturity would also be interesting.

Snobar:
Does the harvest index of a crop vary according to whether it is irrigated or rainfed?

Capper:
Irrigated crops tend to have higher harvest indices but rainfed crops tend to have better digestibility because of incomplete translocation of carbohydrates to the grain in many instances.

Abd El Moneim:
You referred to the toxic components in straw and grains such as phenols, tannin, alkaloids, etc. From the breeders' point of view there is a positive correlation between these and resistance to disease and pests. Animal nutritionists should advise the breeder about the acceptable levels of these components, and should not ask for varieties completely free from them, or crops will be completely devastated by disease or pests.

Capper:
It is not at all clear how important tannins are in food legume straws as they occur mainly in the seed coat, but it is becoming apparent that they are not necessarily responsible for chemical defence against pests and disease. They may be formed in response to stress such as water shortage. Tannins may protect protein from rumen degradation so they may be beneficial rather than harmful. There is a negative correlation between digestible energy value of some legume straws and lignin content, but it may not be possible to reduce lignin content without affecting lodging resistance. It should be mentioned that in cereal straws no correlation has been demonstrated between improved straw quality and lodging resistance. I am suggesting that it may be useful to determine what the various phenolic compounds are and how they affect feeding value rather than breed for their elimination in the first instance.

PART III

THE ROLE OF LEGUMES IN CROP-LIVESTOCK SYSTEMS

Farming Systems Producing Livestock in Mediterranean Areas

A.R. ABOU AKKADA
Faculty of Agriculture, University of Alexandria, Alexandria, Egypt

Abstract. The major farming systems producing livestock in Mediterranean areas can be classified into four groups: nomadic herding; transhumant farming; rainfed farming; and irrigated farming. In these systems the animal resources and the feed resources closely interact. In a study of seven Mediterranean countries it was found that small ruminants account for 25% of the total ruminant population, and that natural rangeland and dry roughages account for 75% of the total feed resources, while large ruminants could only be fed by the irrigated farming systems of Morocco and Egypt. The farming systems in these seven countries have generated three livestock feeding systems: extensive; integrated with arable crops; intensive. In mixed farming systems interactions between crops and livestock often have major impact on the productivity and efficiency of these systems.

A detailed four-point strategy is suggested and discussed for the development of the animal feed resources in Mediterranean areas:

1. Programs for the development of natural rangelands.
2. Better use of crop residues and agro-industrial by-products.
3. Increased forage production.
4. The use of forage trees in ruminant feeding.

Introduction

A system comprises interrelated and interacting components (or sub-systems). The operational units of agriculture may be described as agricultural sub-systems, including all the variations in size and complexity. There are many biological sub-systems within agriculture: every animal and every crop can be regarded as a separate sub-system. However, the importance of these components in agriculture rests on their interaction with each other. Studying isolated components without regard to these interactions will not necessarily lead to agricultural development: no improvement in animal production, for example, can be achieved without considering the farming systems producing livestock.

It is essential to classify the farming systems because it is not possible to cope with the many thousands of individual systems that currently exist

A.E. Osman et al. (eds.), The Role of Legumes in the Farming Systems of the Mediterranean Areas, 165–178.

(Spedding 1979). Thus, unless we can generalize about categories such as livestock-production systems or arable farming systems, we shall not be able to discuss them usefully, nor plan them, carry out research into them, legislate for them nor even name them. We therefore need to classify individual systems into groups. There are several different ways of doing this with farming systems. Ruthenburg (1971) has described the farming systems of the tropics and discussed the main ways in which systems of cultivation may be classified. He classified livestock farming according to the degree of migration of both the animals and their owners. His main categories are:

1. Total nomadism
2. Semi-nomadism
3. Transhumance
4. Partial nomadism
5. Stationary animal husbandry

Duckham and Masefield (1970) have suggested that the factors determining the location of farming systems are:

1. Climate: precipitation and evapo-transpiration
2. Land: topography, bio-gas chemical properties and soil stability
3. Moisture control facilities: irrigation and drainage
4. Unwanted species: pests, disease and weeds
5. Operational facilities
6. Availability of inputs
7. Availability of markets
8. Feasible crops and animals

Thus, the farming systems in the Mediterranean areas with hot, dry summers and cool wet winters and springs, should be different from those in the temperate regions. Similar considerations apply to livestock production sub-systems. They depend upon the crops or vegetation that can be grown, and are further influenced by climate; by the nature and quantity of feed available; and by the incidence of pests, parasites and disease.

Farming Systems Producing Livestock in the Mediterranean Areas

The range of possible classification schemes really reflects the variety of ways there are of looking at the farming systems in Mediterranean areas. There are, however, some common features that are encountered in any classification. For the purpose of describing livestock production activities, four main types of farming systems can be identified:

1. Nomadic herding
2. Transhumant farming
3. Rainfed farming
4. Irrigated farming

Nomadic Herding

Nomads mainly produce livestock – cattle, sheep, goats and camels. They are characterized by continuous movement along fixed routes year-round in search of pastures. They are seldom cultivators but they obtain the cereal they need through barter. Nomadic husbandry is confined to the arid and semi-arid regions. Nomadic herders are often viewed as a menace to rangelands and accused of defoliation and deforestation. The true extent of their impact on rangelands needs research and quantification. There is reason to believe that in this system, manure provided by livestock slows the deterioration of rangelands.

The precarious nature of nomadic existence induces nomadic people to hedge against uncertainties by carrying excess stock. Their wandering nature precludes the adoption of new technologies for reducing risk and thereby reducing flock numbers without any loss of total output. Strategies should be set up for expanding pastures and ranges, for providing more watering points, and for advice through extension on limiting stock numbers to prevent overgrazing.

Transhumant Farming

Transhumant agriculturists are sedentary and livestock constitutes their major on-farm resource. They live in arid and semi-arid regions, travelling to higher pastures during the summer and returning to their homesteads in the winter. They grow staples such as wheat, corn and other cereal grains. Strategies should be formulated in order to improve breeds and introduce irrigation for higher and more stable yields.

Rainfed Farming

Rainfed agriculturists are sedentary and are found in the arid and semi-arid regions. Livestock, especially small ruminants, play an important role in rainfed agriculture. Rainfed agriculturists depend mainly on cereal cultivation which is weather dependent, and so outputs tend to fluctuate unpredictably. Strategies for development of animal production in rainfed areas should focus on introducing crop rotations and intercropping of legumes to enlarge the feed base; on improvement of animal quality; and on promotion of more efficient use of by-products and their preservation.

Irrigated Farming

Irrigated farming is a means of intensifying production through controlling water. This system is used worldwide and is common in Mediterranean areas. The livestock usually used in irrigated agriculture are large ruminants. They draw water from wells, provide farm traction and supply manure for crop cultivation. Irrigated agriculturists are intensive cereal cultivators and are sedentary. Strategies for livestock production should include legume intercropping; improvement of animal quality; and promotion of more efficient use of by-products and their conservation.

Effect of Farming Systems on Livestock Production in Mediterranean Areas

Farming systems may influence the environment in several ways. They may change the vegetation and appearance of the landscape, and exert an influence on man and animals by their manner of operation. The most important ways in which they influence livestock in the Mediterranean areas are:

Animal Resources

Table 1 shows the distribution of ruminants in seven Arab countries in the Mediterranean areas. The small ruminants predominate: sheep and goats in Mediterranean countries form 64.19% and 20.36% of the total population of ruminants in the seven countries, respectively. The ratio of sheep to goats is approximately 3:1. Algeria and Morocco have the largest goat and sheep population, accounting for 58.0% of the total population of the small ruminants. The sparse vegetation provides a weak feed resource base which can sustain only sheep and goats. For this reason, the small ruminant production systems in the seven countries are primarily nomadic and transhumant. Sheep and goats fit well into the farming systems, their greatest value being their small size, rapid turnover, and the conversion of feed resources not directly eaten by man: i.e. natural rangeland, fallow grazing, browse, and crop residues.

Buffalo and cattle account for only 15.44% of the total population of the ruminant resources in the Mediterranean areas (Table 1). Buffalo are kept only in Egypt and cattle are predominantly kept in Egypt and Morocco, where the irrigated farming systems provide significant amounts of green

Table 1. Animal resources in seven countries in Mediterranean areas.

	Animal Populations (millions)			
Countries	Buffalo	Cattle	Sheep	Goats
Syria		0.73	7.58	1.10
Lebanon		0.10	0.23	0.34
Egypt	2.44	2.70	2.45	2.37
Libya		0.18	4.84	1.68
Tunisia		0.92	4.97	0.92
Algeria		1.35	12.32	3.82
Morocco		3.15	15.71	6.03
Total	2.44	9.13	48.10	16.26

Source: Abou Akkada et al. (1984).

fodders suitable for their development. Because of the abundance of irrigated forages, sedentary systems that are also intensive systems have become possible. When grazing is conducted on the green fodder, buffalo and cattle tend to be grazed separately, but when sheep and goats are reared, these small ruminants are grazed together.

Data in Table 1 clearly indicate that ruminant production systems form a component of the farming system prevailing in the Mediterranean areas. The choice of the ruminant suitable to any one of these coutries has been evolved in response to the total availability of land, the type of crop production practised, the frequency of cropping, the area of marginal land, and the amount of vegetation.

Feed Resources and Feeding Systems

Feed resources available in the seven countries are included in Table 2. Natural rangeland and dry roughages account for 37.7% and 36.1% respectively of the total feed dry matter produced there. The greatest proportion of natural rangeland is in Algeria and Morocco. The dry roughages consist of crop residues such as rice and wheat straws; by-products of vegetable and fruit and sugar-cane processing; and stover from arable cropping. Both natural rangeland and dry roughages are contributed by rainfed and irrigated farming systems prevailing in the selected countries. In these two farming systems, the production of the natural rangeland and the roughages does not compete with the growing of crops.

The green forages account for 20% of the total dry matter resources. Almost all the green forages are produced in the irrigated farming system prevailing in Egypt. Berseem (*Trifolium alexandrinum*) is considered to be the most important winter legume forage on which livestock feeding depends, especially for milk and meat production. The annual cultivated area

Table 2. Animal feed resources in seven Mediterranean countries (million tonnes/year).

Countries	Natural rangeland (Dry matter)	Irrigated fodders		Dry roughages	Concentrates
		Fresh matter	Dry matter		
Syria	2.00	0.67	0.13	1.72	1.05
Lebanon	0.05	0.02	0.003	0.02	
Egypt	0.36	59.00	10.620	12.80	2.59
Libya	1.67	1.08	0.195	0.34	
Tunisia	2.48	3.24	0.583	2.15	0.17
Algeria	8.41	2.70	0.485	3.72	0.40
Morocco	12.63	9.41	1.494	5.61	1.13
Total	27.60	76.12	13.838	26.36	5.34

Source: Abou Akkada et al. (1984).

of berseem is about 2.8 million feddans (1.1 million ha). The area cultivated with alfalfa is 22 000 feddans (8800 ha).

As a result of the influence of the farming systems on the quality and quantity of feed resources, three feeding systems have been generated in the Mediterranean areas:

1. Extensive Feeding System
This is the most common. It is characterized by ruminants, usually owned by smallholders, grazing on all available grazing areas, including rangeland, for varying periods during the day. More animals tend to be herded than the carrying capacity of the pasture, thus causing overgrazing and deterioration of the rangeland. In this feeding system, animals are maintained on a low level of nutrition where very few or no supplements such as concentrates, salt or mineral licks are provided.

2. Feeding Systems Integrated with Arable Cropping
These have been evolved in the situation where a large proportion of the livestock feed is derived from the arable cropping stubbles and residues. The animals may also graze grasses and browse shrubs found in the fields. Animals, therefore, do not compete for the same land and may play a supplementary role to arable cropping in the rainfed or irrigated farming systems.

3. Intensive Feeding Systems
Here large concentrations of animals are housed most of the time (stall feeding). The animals depend mainly on high-energy concentrate feeds with crop residues as the roughage proportion of the ration. In some localities, food processing and agro-industrial by-products such as rice and wheat brans, cottonseed cake and soybean meal provide the basis for intensive ruminant production systems. Intensive feeding may also be based on crop residues and good quality cultivated fodders such as berseem or alfalfa. Intensive feeding systems occur primarily in integrated crop-livestock farming systems and in situations where there are abundant supplies of agro-industrial feeds.

Crop-Livestock Integration

An integrated project on desert agriculture has been jointly conducted by the University of Alexandria and the American University, Cairo (AUC) in the desert of South Tahrir, Egypt (Abou Akkada 1988). The main objectives of this project are: i) to describe the bio-economic aspects of the many crop and livestock activities of desert farming systems; ii) to develop appropriate farming systems in the desert; and iii) to provide training material and resource management in desert farming systems.

Three model farms were established in the desert of South Tahrir, Egypt. Details of the cropping patterns are included in Table 3. The economic

Table 3. Profitability of the cropping patterns of model farms.

	Cropping Pattern	Area (ha)	Net Returns (L.E.)[1]	
			Per Farm	Per Hectare
Farm 1 (8.4 ha)	Alfalfa	6.3		
	Perennial forage grass	2.1	2960	355
Farm 2 (7.5 ha)	Alfalfa	4.2		
	Medicinal plants (roselle + rehan)	0.8		
	Annual crops (barley)	1.7	3540	473
	Vegetables (peas + onions)	0.8		
Farm 3 (8.4 ha)	Alfalfa	1.7		
	Annual fodder crop (berseem)	2.5		
			2870	429
	Annual field crop (lupins + peanuts)	2.5		
	Fruit[2]	1.7		

[1] L.E. = 0.44 USD approx.
[2] Area allocated for fruits was committed to peanut production in years 1 and 2 of the project.

Source: Abou Akkada (1988).

analysis of the alternative farming systems is also shown. Based on the collective contribution margins for each cropping pattern in the three model farms, Farm 2 was found to have the highest net returns per hectare (L.E. 473). This is followed by Farm 3, with a positive contribution margin of L.E. 429 per hectare (L.E. = 0.44 USD approx.). It is therefore indicated that the farming system adopted in Farm 2 in the first year of the project is the most efficient with respect to profitability. It should be emphasized that this analysis is for crop production only, without taking the livestock component into consideration.

Table 4 shows the net returns of the three model farms from both crop and livestock production. The results indicated that Farm 2 again had the highest net returns after the livestock production was included. Animal production generated positive net returns for Farm 1, offsetting the lower net returns resulting from crop production alone (Table 3).

Results in Tables 3 and 4 show that in mixed farms, i.e. those with both crop and livestock components, interactions between these components often have major impact on the productivity and efficiency of these systems. Some interactions are indirect, resulting from competition between crop and livestock. Complementary interactions also occur. Relatively little research has been done on the interactions among the soil, crop and livestock components of mixed farming systems. Such research could be oriented towards improving crop-livestock production systems.

Table 4. Profitability of cropping patterns and livestock production for model farms.

	Cropping pattern	Area (ha)	Livestock		Net returns per ha.(L.E.)
			Type	No. of heads	
Farm 1	Alfalfa	6.3			
			Cow	14	616
	Perennial grass	2.1			
Farm 2	Alfalfa	4.2			
	Medicinal plants	0.8			
			Sheep	64	727
	Annual crops	1.7			
	Vegetables	0.8			
Farm 3	Alfalfa	1.7			
			Sheep	12	338
	Annual fodder crops	2.5			
	Annual field crops	2.5	Cow	2	
	Fruits	1.7			

Source: Abou Akkada (1988).

Strategies for the Development of Livestock Feed Resources in Mediterranean Farming Systems

Feed resources represent the overriding constraint in the livestock production systems in the Mediterranean regions. Thus, any development in livestock production can only be achieved through the improvement of animal feeds, by the following nutritional strategies:

Programs for Development of Natural Rangelands

Forages from natural rangelands provide a major component of the animal feed resources in Mediterranean areas. However, in recent years, pressure from the increasing animal populations has led to overstocking and overgrazing. One solution to these problems is to improve the productivity of the natural rangelands. Abou Akkada et al. (1984) have discussed in detail the following strategies for the development of natural rangelands in countries of Mediterranean areas:

a. Programs for rain water harvesting and distribution in the semi-arid and arid regions.
b. The appropriate distribution of water points as a means of control of grazing activities.
c. The establishment of feed reserves to be utilized during the dry periods where vegetation is scarce.
d. The establishment of range reserves in order to secure an abundance of fodders during the critical periods.

e. The establishment of livestock cooperatives in order to secure the cooperation of the livestock owners in introducing range protection and conservation practices.
f. Range revegetation programs in the overgrazed areas of potential productivity.

Better Use of Crop Residues and Agro-industrial By-products

Not enough use is being made at present of the large quantities of crop residues and agro-industrial by-products to feed ruminants (Abou Akkada et al. 1984). This is possibly due to an inadequate use of intensive feeding systems. As population pressure increases and the area devoted to food-crop production is extended, the use of crop residues and by-products for ruminant feeding should also increase.

Most of the agro-industrial by-products, such as vegetable and fruit processing by-products, can be used in feeding the ruminant without any treatment (Abou Akkada 1986). However, the productivity of animals fed crop residues such as straws, cobs and stovers can be increased by chemical treatment (e.g. ammonia and caustic soda) of the residues to increase digestibility and feed intake. In the Mediterranean countries, the major problem with this approach is to find methods that are simple, effective and acceptable at the village level.

There are a number of ways in which ammonia can be used to increase the nutritive value of fibrous crop residues, including the use of ammonia gas, ammonia in solution or ammonia generated from urea. For small farmers, it is more convenient to generate ammonia from urea by the 'wet ensiling process'. Urea is a common fertilizer which is often subsidized and which farmers have become accustomed to handling. Ammonia is generated more rapidly at higher temperatures, which makes the system appropriate for countries of the Mediterranean region.

In Egypt, reconnaissance surveys indicated that the annual amounts of crop residues and agro-industrial by-products such as wheat and rice straws, maize (corn) stover, rice hulls and by-products of vegetable and fruit processing are in the order of 16 million tonnes (Abou Akkada 1984). Large scale on-farm trials were conducted through a national program on the utilization of by-products as animal feeds, to study the use of urea-treated and untreated rice straws in cattle finishing in the rural areas of Giza Governorate (Shinnawy and Abou Akkada 1987). Based on the biological and economic data (Table 5), the results clearly demonstrated the superiority of the urea-treated rice straw-based ration over that containing untreated rice straw. The response of the farmers to the rice straw-based ration was positive. Experiments in Alexandria University have also indicated that rice hulls (Abou Akkada and Nour 1985), corn stover, and by-products of vegetables and fruit processing (Nour et al. 1980) can be used in ruminant feeding.

Table 5. Performance of cattle finished[1] on rice straw-based rations in rural areas of Egypt.

	Untreated straw (Control)	Urea-treated straw
No. of farmers	13	26
No. of animals	50	100
Finishing period (days)	180	180
Liveweight gain (kg)	175	197
Cost of feeding (L.E./kg of bodyweight gain)	2.66	2.02
Total cost of finishing (L.E.)	1319.17	1251.76
Net returns (L.E.)	294.83	428.24
Economic efficiency %	116.00	150.00

[1] Cattle in all treatments received conventional concentrates in addition to the rice straw.

Source: Shinnawy and Abou Akkada (1987).

The most important by-products that offer good potential as animal feeds in the Mediterranean areas are straws (wheat, barley and rice), maize (corn) stover and cobs, sugar-beet pulp, sugar-cane by-products (tops, molasses, bagasse and pith), and hulls (cottonseed and rice), because of:

a. the large quantities available but unutilized.
b. the absence of conflict in their use with human food needs, and
c. the potential for improving their feed value by chemicals.

Increased Forage Production

The need to commit a major proportion of land to the production of food crops limits the area that can be used for forage crops in the countries of Mediterranean regions. However, attempts should be made to increase cultivation of forages (grasses and legumes) on available land, which includes uncultivated or newly reclaimed land. Forage crops can also be incorporated in rotations with food crops. Forage legumes are very often planted to restore soil fertility before a food crop is planted.

Berseem (*Trifolium alexandrinum*), alfalfa, cow-peas, maize forage (Darawa), elephant grass, sorghum and Sudan grass are the most widely used forages for feeding ruminants in Mediterranean areas. The presence of such forage reserves forms an important component of integrated agriculture in small farms. The cultivation of improved species such as Sordan (Sudan grass x Sorghum hybrids) may offer an excellent opportunity for increased fodder production (Soliman and Abou Akkada 1987). The introduction of forage legumes and grasses to Mediterranean conditions is of prime importance for the balance of animal rations and better crop rotations. Seeds of three winter forage legumes; *Vicia sativa*, *Vicia dasycarpa*, *Lupinus albus* (forage lupins); and three summer forage crops: *Lablab purpureus*, var. Highworth, and Guar, and Rhodesgrass (*Chloris gayana* Kunth) were tested under the desert conditions of South Tahrir, (Abou

Akkada 1988). Results of forage yields and seed production indicated good potential as additional sources of higher quality animal feeds.

Greater opportunities exist for increased forage production in growing good grass-legume mixtures matching pasture growth with animal requirements. In any good grass–legume pasture, special attention must be given to establishing the legume and ensuring its production. Effective nodulation is the first prerequisite for good grass–legume pastures. On-farm trials have been carried out on 87 farms in 27 villages of Behera Governorate, Egypt, in collaboration with 467 smallholders, in order to evaluate the productivity of barley-berseem mixtures (Soliman and Abou Akkada 1988).

Results in Table 6 clearly indicate the superiority of barley-berseem mixtures to berseem alone in forage production. The greatest improvement in forage yield is in the first, second and third cuts; thereafter it declines. Both berseem-barley and berseem were inoculated with a proven *Rhizobium* strain.

The basic strategy is to produce sufficient amounts of good-quality feed all year round. Thus, innovative methods are needed for enabling seasonal surpluses of forages to be preserved for subsequent use when feeds are in short supply, such as during critical periods and dry seasons. Such methods should be simple and within the capacity of smallholders. Hay-making is the most common procedure for conserving the feed surplus. Silage-making offers good potential for feed preservation in situations where hay making is not possible. Table 7 shows the performance of cattle fed on silage made from green fodders and crop residues. The nutritive value of silage is in some cases comparable to the conventional concentrate feeds commonly used in Egypt (Soliman and Abou Akkada 1987). The response of small farmers to silage was positive because of its value as a feed reserve in the critical feeding periods in the summer season. However, there is a widespread lack of information on the actual utilization of feed resources in silage making in the existing cropping systems.

Table 6. Forage production of barley-berseem mixture in rural areas of Egypt.

Cuts	Fresh yield (t)		Increase over berseem monoculture (%)
	Berseem	Berseem – barley mixture	
1st	10.5	14.1	34.3
2nd	12.7	15.5	22.0
3rd	12.0	14.1	17.5
4th	11.9	13.5	13.4
5th	5.1	5.6	9.8

Source: Soliman and Abou Akkada (1988).

Table 7. Performance of cattle fed on silage based on green fodders and crop residues in rural areas in Egypt.

	Nutritive value (TDN%)	Milk yield (kg/day)
Berseem (2nd cut) + faba bean stalks + molasses	52.0	5.95
Berseem (1st cut) + green barley (2:1)	54.8	6.02
Berseem (2nd cut) + corn stover (2:1)	52.0	5.48
Berseem (2nd cut) + rice hulls (20%) + molasses	57.4	NA
Green corn stover (95 days after planting)	63.0	6.61
Corn fodder (stalks + ears)	64.2	NA

Source: Soliman and Abou Akkada (1987).

The Use of Forage Trees in Ruminant Feeding

In recent years, there has been a growing awareness world-wide of the general multipurpose value of forage trees and their specific role in cropping-livestock systems. However, farming of fodder shrubs is virtually unknown in many Mediterranean countries. Many species of *Acacia* have proved to be useful multipurpose trees in North Africa (El-Lakany 1987). Some are browsed regularly and play a significant role as a reserve and supplementary fodder. *Acacia saligna* may survive and grow on sites receiving as little as 200 mm of rain annually, or even less.

Acacia saligna has been the subject of several studies which dealt with fodder production, palatability, intake and digestibility in the University of Alexandria and American University (AUC) project on desert agriculture (Abou Akkada and Bishay 1986), in south Tahrir, Egypt. The shrubs were grown and samples taken for proximate analysis. Two digestibility trials were carried out on Barki lambs. In the first trial, animals were fed on alfalfa as the basic ration, while in the second, *Acacia* foliage was offered to the animals with alfalfa. The animals were weighed regularly throughout the two experiments. Average digestion coefficients and nutritive values of the rations were evaluated. The results of the digestibility trials are included in Table 8. It is noted that introducing *Acacia* to the lambs' ration has resulted in an increase in the dry matter intake by the lambs and in the nutritive value of the ration. However, supplementing alfalfa with *Acacia* has decreased the digestible protein (DP), thus saving the loss in alfalfa protein when alfalfa fodder is the sole ration.

Research should be conducted to study the effects of feeding forage trees

Table 8. Effect of supplementing alfalfa with *Acacia* for feeding lambs.

Ration	Dry Matter Intake kg/day	% of body weight	Nutritive value (TDN%)	Digestible protein (DP)
Alfalfa	1.058	2.35	59.1	12.5
Alfalfa + *Acacia*	1.648	3.66	64.2	8.64

Source: Abou Akkada and Bishay (1986).

and shrubs on the productivity and reproductive performance of ruminants. Economic evaluation of this feeding system and the response of farmers to its application should also be undertaken in countries in Mediterranean areas. The specific role of forage trees is only as a supplement to crop residues: they should not be considered as forming the whole diet. Furthermore, growing fodder shrubs and trees is usually confined to land not suitable for crop production. Under these circumstances, forage shrubs and trees may play a significant role in ruminant feeding in desert farming systems.

References

Abou Akkada, A.R. 1984. Some prospects for the development of feed resources in Egypt. Proceedings of Second Symposium of Feed Manufacturing and Quality Control. Ministry of Agriculture and U.S. Feed Grains Council, 27–28 October 1984, Cairo, Egypt.

Abou Akkada, A.R. 1986. By-products as future animal feeds in developing countries. Proceedings of the Second Egyptian-British Conference on Animal and Poultry Production, 26–28 August. Bangor, Wales, U.K.

Abou Akkada, A.R. 1988. Applied research, demonstration and training in desert development: Proceeding of Egyptian-American Workshop on Agriculture, Supreme Council of Universities, 9–11 January 1988, Cairo, Egypt.

Abou Akkada, A.R. and Bishay, A. 1986. Ninth Technical Report: Applied research, demonstration and training in desert development. Supreme Council of Universities. Cairo, Egypt.

Abou Akkada, A.R., Farid, M., Wardah, M., Hassan, N., Al-Shorbagy, M., Bayoumi, M., and Alwash, A. 1984. Evaluation of present status and potential development of animal feed resources in Arab Countries; the national study. Arab Organization for Agricultural Development (AOAD), Khartoum and the Arab Centre for the Studies of Arid Zones and Dry Lands (ACSAD), Damascus, ACSAD/AS/P53/1984. In Arabic.

Abou Akkada, A.R. and Nour, A.M. 1985. By-products utilization in Egypt. Proceedings of Second Workshop of 'ARNAB' on Optimal Feeding of Agricultural By-products to Livestock in Africa, 14–18 October 1985. Alexandria, Egypt.

Duckham, A.N. and Masefield, G.B. 1970. Farming Systems of the World. Chatto and Windus, London.

El-Lakany, H.H. 1987. Protective and productive tree plantations for desert development. Proceedings of 2nd International Conference on Desert Development, 25–31 January 1987, Cairo, Egypt.

Nour, A.M., El-Shazly, K., Abou Akkada, A.R., Borhami, B.E. and Abaza, M.A. 1980. Evaluation of silage of some by-products from food processing industry. Alexandria Journal of Agricultural Research 18:1.

Ruthenberg, H. 1971. Farming Systems in the Tropics. Oxford University Press, Oxford, UK.
Shinnawy, M. and Abou Akkada, A.R. 1987. Annual Technical Report: National Research Program on Animal Feeds. National Academy of Science, Cairo, Egypt.
Soliman, S.M. and Abou Akkada, A.R. 1987. Annual Technical Report: National Research Program on Animal Feeds. National Academy of Science, Cairo, Egypt.
Soliman, S.M. and Abou Akkada, A.R. 1988. Annual Technical Report: Integrated Rural Development Project, Behera Governorates. Ministry of Agriculture, Cairo, Egypt.
Spedding, C.R.W. 1979. An Introduction to Agricultural Systems. Applied Science Publishers Ltd., London, U.K.

Discussion

Capper:
I would like to comment on the effectiveness of urea treatment of straws. It has been shown that urea treatment does not bring about splitting of ligno-cellulosic bands as the pH does not get high enough. This may be caused by the formation of ammonium carbonate through reaction with carbon dioxide in the air. Would it not be better to feed the urea directly to the animals rather than go through the process of ensiling straws with urea, which is time-consuming and of limited effectiveness?

Akkada:
Various researchers have clearly indicated that cellulose digestion was improved when poor quality roughages were treated with ammonia and urea. I personally do not favour the feeding of urea directly by smallholders because of possible hazards of mishandling. Efforts are being made to feed urea in the form of urea-mineral-molasses blocks.

Papastylianou:
What is the percentage of animals in WANA in the Nomadic and Transhumant groups?

Akkada:
In the AOAD/ACSAD study we found that about 73% of the total animal population is maintained on natural rangeland, so we could conclude that 73% of the animals belong to the nomadic groups.

The French Mediterranean Zones: Sheep Rearing Systems and the Present and Potential Role of Pasture Legumes

G. GINTZBURGER,[1] J.J. ROCHON[2] and A.P. CONESA[1]

[1]*INRA-Lecsa, 9 Place Viala, 34060 Montpellier, France;* [2]*Agronomie-IUT, Perpignan, France*

Abstract. Once an agriculturally active part of France, the French Mediterranean regions are struggling with changing agriculture and economics within Europe. Concentrated in the coastal zone, wine industries led local agriculture for nearly a century but are currently facing hardship from excess production. Sheep rearing, a relict of deceased farming systems, is the dominant livestock industry which is also suffering from low productivity and rural decline in the hills and mountains of the region. The past, present and future role of forage legume resources within farming systems are briefly reviewed. While perennial forage legumes are well known and locally used, the potential of annual forage legumes such as subclovers and annual medics has still to be explored for both intensive and extensive sheep rearing systems of the South East of France.

Introduction

The French Mediterranean zone covers some 6.8 million hectares approximately 12% of France (SCEES 1983). It is sheep country by tradition because of the essential role once played by this animal in local farming systems. At present, there are in this zone four to six heads of sheep to one of cattle, compared to a 2:1 ratio of sheep to cattle in France as a whole. It is also a country of large tracts of rough uncultivated land, once largely contributing to feeding large flocks of sheep. As for the acreage of fodder resources, nearly 65% of the south east of France is covered with rangeland, while 23% is considered as permanent pasture, and only 13% is registered as annual sown pasture, compared to 12, 52, and 30%, respectively, for France as a whole (Mansat 1980). The sheep industry in the south is surviving with extreme difficulty in spite of its recognised role in assisting control of bush encroachment, the latter being responsible for the large extent of dramatic bush fires every summer.

In this paper, we present a brief account of land use and agriculture in the French Mediterranean zones with the past, present and future role played by sheep rearing there. We shall then discuss the potential place and role of pasture legumes within grazing systems.

A.E. Osman et al. (eds.), The Role of Legumes in the Farming Systems of the Mediterranean Areas, 179–194.

Geography, Climate, Soils and Natural Vegetation

The northern limit of the French Mediterranean zones is sometimes delineated by the northern extent of olive trees. This is in fact too restrictive as olive trees are frost-sensitive and disappear in cold environments still subjected to Mediterranean influence, i.e. rainy winters, dry summers. Climatically speaking, a better definition remains with 'All the area receiving mainly winter rains, with a high level of intra- and inter-annual rainfall variability'. Therefore, the French Mediterranean zone is not strictly restricted to the coastal fringe but also includes some hilly regions, cold plateaux and mountains where occasional oceanic and continental summer influences occur (Table 1). Taking this into account, one can identify a large strip 50–120 km wide along the Mediterranean coast. It includes three zones along a south-north transect from the Mediterranean Sea to the mountains, which are:

a) a densely populated coastal zone with a strict Mediterranean climate, where most of the present agricultural activities are concentrated, such as intensive vineyards, fruit trees, vegetables and flower growing with little or occasional animal production.
b) a hilly 'dry' country with bushes and small trees where there is isolated mixed cropping including some cereals and vineyards, but where sheep rearing is significant because of large tracts of rangeland. This zone enjoys a predominantly Mediterranean climate, though colder than a), and receiving more rainfall.
c) a forest and rangeland high country of plateaux and mountains, mostly devoted to sheep and cattle rearing. The climate is colder than in the other two zones. Some strong continental and oceanic influences induce more reliable and higher rainfall throughout the year (Fig. 1).

Table 1. Climatological aspects of the three administrative regions of the French Mediterranean zone.

	Min °C	Max °C	Average annual rainfall	Days of frost	Elevation (m)
Languedoc					
a) Montpellier	1.8	28.3	690	35	40
b) St-Martin-de-Londres	−0.2	28.3	1160	66	200
c) Le Caylar	−2.1	25.0	1240	110	730
Provence					
a) Marignane	2.3	29.0	547	30	25
b) Aix-en-Provence	0.5	26.8	608	58	173
c) Embrun	−3.6	26.0	1090	104	800
Corse					
a) Ajaccio	3.9	27.6	660	8	4
b) Ocana	3.6	30.1	996	9	280
c) Bastelica	−0.9	27.4	1354	82	800

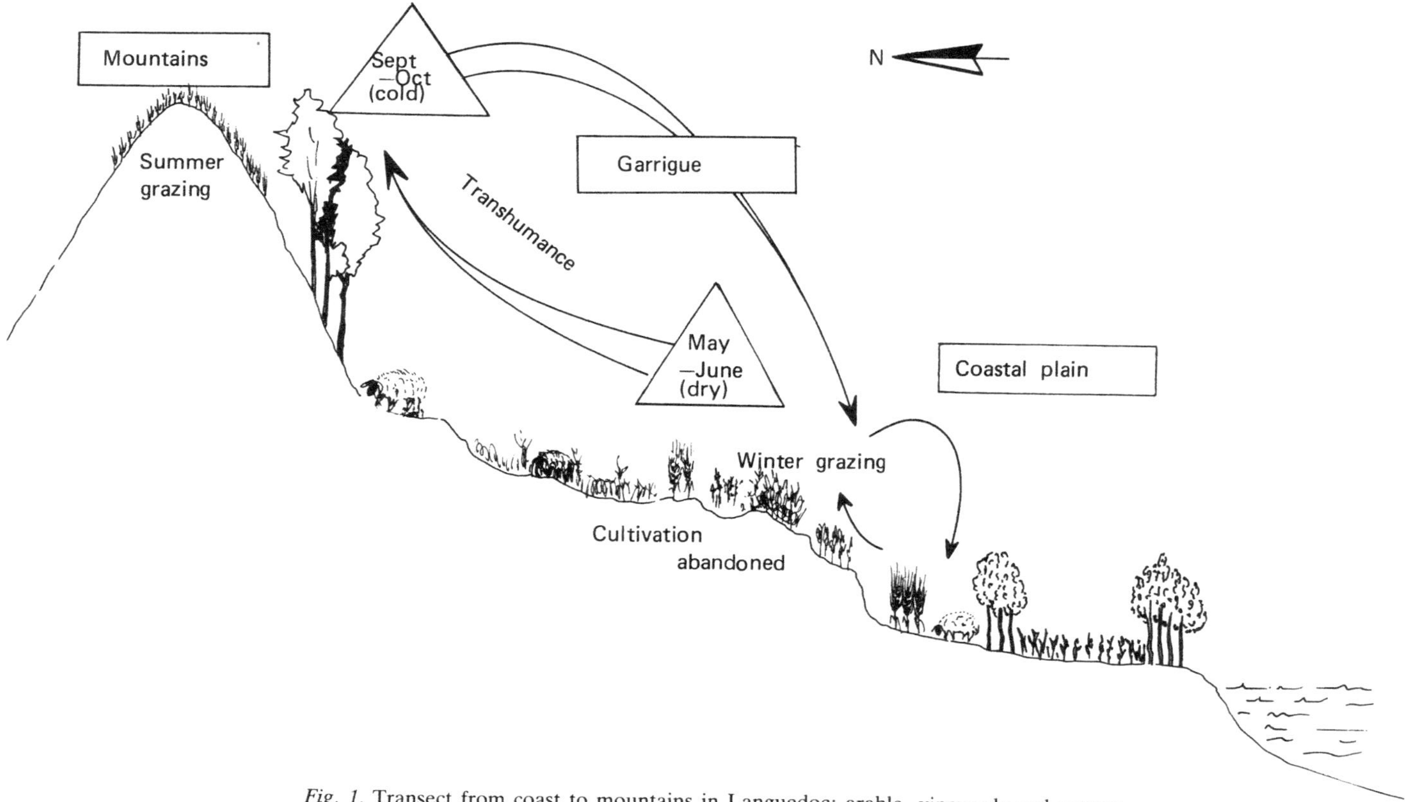

Fig. 1. Transect from coast to mountains in Languedoc; arable, vineyards and pasture.

This physiographical, and agricultural, organization of the landscape can be observed in the three administrative regions of the French Mediterranean zone which are:

1. The Languedoc-Roussillon, west of the Rhone valley and squeezed between the southern slope of the Massif Central and the Mediterranean Sea to the Pyrenees Mountains on the Spanish border. Westward, the Roussillon has definitely acid soils. Oak trees (*Quercus suber*) are common in the hills covered with a dense bush of *Erica* sp. and *Arbutus unedo*. Chestnuts (*Castanea sativa*) occur in dense stands at higher altitudes.

   Languedoc has predominantly calcareous soils. In the hills (100–500 m elevation), *Quercus ilex* and *Quercus coccifera* form a dense low scrub interspersed with abandoned fields, well-maintained vineyards cleared of stones, and plantations of *Pinus halepensis*. A short grass (*Brachypodium ramosum*) mixed with woody *Thymus* sp. or *Rosmarinus officinalis* spreads over open rangeland subjected to bush encroachment. All this country is called 'Garrigue' locally.

   At higher altitudes, open forests of white oak (*Quercus lanuginosa*) intermix with rangeland covered with bush, Box (*Buxus sempervirens*) and dotted with *Brachypodium* sp.
2. Provence-Cote d'Azur, spreading east of the Rhone, to the Italian border and backed by the Alps. Most of Provence is calcareous, while some (Maures and Esterel mountains) is acid. A landscape similar to 'Garrigue' with omnipresent olive trees, *Thymus* sp., and *Rosmarinus* sp. dominates with plantations of *Pinus pinea* and fields of *Lavandula latifolia*.
3. Corsica is the third region, off the French-Italian coast. Acid soils are dominant and covered with a very dense bush, a type of vegetation similar to the one encountered in the Roussillon, known here as 'Maquis'.

A Changing Agriculture

For historical reasons (Boutonnet 1981), agricultural activities in the French Mediterranean zone have been seriously affected. This trend, accelerating in the 1950s, has hit the once densely populated and agriculturally very active French Mediterranean zone very hard.

Several reasons were responsible: the progress in agriculture in northern Europe, better means of communication combined with refrigerated transport, a decline in agricultural activities and the strong rural exodus which took place in several regions of France. Rainfed production of cereals, olive oil, grain legumes (chickpeas, lentils, various beans), once widely practised, and the sheep industry, have dropped drastically, to be replaced, since the beginning of the 20th century, by vine growers concentrating on the coastal plain. They now represent some 70, 30 and 20% of agricultural units in Langeudoc-Roussillon, Provence-Cote d'Azur and Corsica regions respectively, compared to an average of 13% in France as a whole (Table 2).

Wood cutting, bush clearing and charcoal industries now obsolete, used to keep large areas of land available for mixed cropping associated with grazing for sheep and goats.

Sheep numbers sharply declined with the rural population. Sheep not only produced meat, milk and wool for local consumption and use, but their manure was used as fertiliser in cropping systems. Concurrently, meat production (predominantly mutton and lamb) geared to fulfill local requirements, and the traditional wool industry, both lost momentum due to high competition from other countries. This trend continues. For example, the Departement de l' Herault (Languedoc) covering 622 700 hectares, had approximately 400 000 head of sheep in 1894, but only 88 000 in 1964 (Loc. cit. CNRS 1985). It has remained stable ever since.

As a consequence, rural migration towards the Mediterranean coastal zone, resulted and is still resulting in large areas of uncultivated and ungrazed areas in the hills. Slow bush encroachment (Fig. 2) feeds large bush fires during the dry Mediterranean summer. The cost of such fires is high and has to be met totally by the local and national community.

More land appears to be becoming available. In fact, all farming systems studies (Cavalier 1983; Barello et al. 1984; Goby et al. 1985) in Languedoc-Roussillon, make it clear that land for grazing is available in excess (Fig. 3).

Table 2. Land use and main agricultural features of the French Mediterranean regions and France as a whole.

Land use	Administrative Regions and Area (1000 ha)			
	Languedoc-Roussillon	Provence-Cote d'Azur	Corsica	France
Cultivated land	261	275	12	17 647
Permanent pasture	508	597	320	12 627
Vineyards & fruit trees	452	184	29	1 370
Abandoned agricultural land	472	331	170	2 742
Forest	796	1104	231	14 593
Non-agricultural and waste land	287	655	109	5 926
Total	2 776	3 180	817	54 908
Wine growers	70%	30%	20%	13%
Sheep industries (in % of agricultural units)	3.3%	14%	36%	n.a.
Head of sheep (1000)	557	1106	132	13 118
Ewes (1000)	347	633	99	8 092
Meat production (mutton and lamb in tonnes)	6 980	12 441	393	195 000

Source: SCEES (1982).

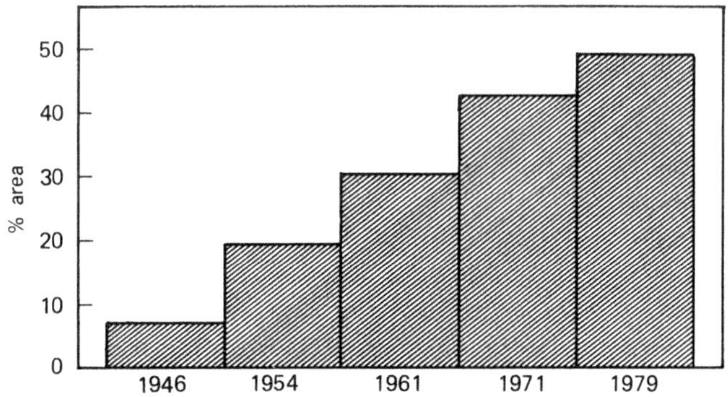

Source of data: CNRS (1985)

* vegetation > 2 m high and > 25% ground cover.

Fig. 2. Trends in forest-bush* encroachment in Languedoc, 1946–79.

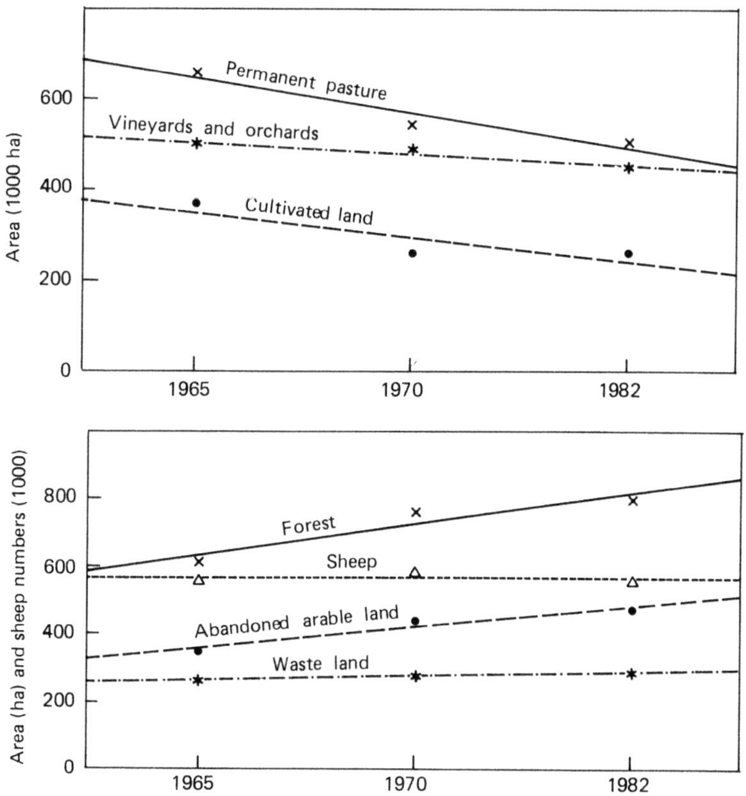

Source of data: SCEES (1983).

Fig. 3. Trends in land use and sheep numbers in Languedoc and Roussillon, 1965–82.

This is a valid assumption for all the French Mediterranean regions. However, several reasons may temper land availability for agricultural activities.

A Frozen Land Tenure System

A major hindrance remains with the local land tenure system where rangeland and uncultivated land is not easily available for livestock industries.

Cavalier (1983), in a study covering seven villages of the Garrigue zone north of Montpellier (Languedoc) stresses the fact that 50% of agricultural units own 2% of the land, while less than 3% own 50% of the land, mostly heath and bush country. One would think that in these conditions, plenty of grazing land would be available and large flocks of sheep could make use of it.

Competition From Urban Development and Hunting

The sunny South of France, however, remains very attractive to land developers and tourist industries and local landowners have developed a 'wait and see' policy. They remain reluctant to agree on written leases to farmers or graziers. Verbal agreement on a year-to-year basis is the common rule between lessor and lessee. In turn, lessees (mainly sheep graziers) are reluctant to invest in bush clearing, reseeding, fertilizing, or putting up fences on pastures or grazing land. The latter when available are rented on a yearly basis at 10–20 FF/ha (2–4 USD/ha). But impossible competition springs from hunting societies willing to pay 100–200 FF/ha/year for exclusive hunting rights, on large tracts of the Garrigue (Prudhomme 1986). Local hunters, predominantly urban dwellers and offspring of the leisure society, tend to prohibit grazing on hunting reserves, occasionally destroy fences to enforce hunting rights, and do not easily tolerate other use of the land (Rouche 1987).

The various sheep rearing systems, especially those supplying the meat industry, are trying to adapt in spite of such socio-economic and enviromental problems.

Sheep Rearing Systems for Meat Production

France imported nearly 50 thousand tonnes of mutton and lamb in 1982 and approximately 100 thousand tonnes in 1987 (Boutonnet, personal communication). Thus, because of national demand for mutton and lamb, we shall only be dealing with meat production of sheep in the south east of France, although milk production is of major importance in the highlands (for example, the blue-veined cheese industry of Roquefort collects milk from an area 150–200 km round the center of production). Corsica is not considered here.

As a general pattern, local meat-orientated sheep industries produce lambs. In order to ensure such production at minimal cost, lambing usually takes place when forage is more plentiful and directly available through grazing, i.e. mainly in spring.

In the plain of Roussillon, Goby et al. (1985) have identified three main systems, relying on shepherding as fenced pastures are fairly rare. Flocks of sheep consist of 150 to 300 ewes per shepherd. The system is organised to produce pen-fattened lambs locally known as 'Agneau de Perpignan'. The producer fattens 15-day old lambs, not yet weaned, with concentrates (cereal grain, 30–60 kg/head) and sells them for slaughtering when 26–27 kg liveweight (16 kg body weight) and 8–9 weeks old.

The three systems based on a combination of forage resources, are:

1. settled shepherd with forage crops (3–4 ha of dual-purpose barley; permanent pastures such as *Dactylis glomerata*, *Lotus* sp. and/or alfalfa; summer forage crop of irrigated sorghum) and no rangeland grazing. Stocking rate is about 3.5–4 ewes/ha. The most difficult period is late summer-early autumn. Lambing occurs between November and February. To improve prolificity, and because good quality pasture and forage are available, ewes are usually cross-bred 'Romanov x Ile de France' and rams are 'Bleu du Maine'.
2. settled shepherd with an area of approximately 40% pasture and forage crop and 60% rangeland or uncultivated land. Stocking rate is about 4–6 ewes/ha. Available pasture remains similar to the previous case. However, the shepherd relies more on uncultivated land (uprooted vineyards, fallow land) and grazing weeds (50 to 90% of biomass is annual *Lolium rigidum* with some perennial grasses such as *Cynodon dactylon*, and *Agropyron repens*) in vineyards, during autumn-winter.
3. transhumant shepherd using only fallow land and rangeland. The stocking rate is extremely low (0.5–1 ewe/ha) as large tracts of land are available. From mid-June to late October-early November, these flocks move (Transhumance) to high mountains (1000–2000 m elevation in the Pyrenees) where they feed on good permanent pasture.

For both 2 and 3 systems, using a fair part of fallow and rangeland, the most suitable sheep are breeds better adapted to bad weather and walking from the plain to the mountains such as 'Lacaune' or 'Blanc du Massif Central'.

In the 'Garrigue' of Languedoc, near Montpellier, Bazin and Chassany (1986) gathered information from 50 sheep graziers.

About half of them own or have a written lease on the land they graze. The other half roam on waste land according to verbal lease renewable on a yearly basis. Average flock size ranges from 200 to 450 ewes according to bush encroachment of the site. It is commonly agreed that a shepherd can handle safely and efficiently only 150 to 200 head of sheep in bush country. This is another limiting factor to the development of sheep industries in this region (Martinand 1985).

Sheep breeds commonly available are 'Caussenardes' for meat production. It is a sheep well adapted to the rough Garrigue environment and long transhumance walking. Land available per flock varies from 300 to 500 ha. Shepherding is the rule even when fences have been erected to isolate large interwoven tracts of dense bush and open ranges (Rouche 1987). Sheep in Garrigue are fed on the natural vegetation including that of fallows and abandoned vineyards during autumn, winter and spring. Little consideration is given to fodder crops such as lucerne for hay making, as farm machinery is either not available or obsolete. Supplementary feeding is mandatory before and at lambing time especially for shepherds not owning land. It consists of a yearly average of 6 kg of concentrates per lamb, 18 kg of cereal grain per ewe, and 50–100 kg of hay during winter when open grazing is interrupted by cold and wet weather. Supplementary feeding may represent up to 60% of total expenses (Paturel et Salmon 1982).

To avoid summer stress, transhumance takes place from June to October. Lambing occurs in early spring (mid January–mid March) to make the best use of cheap feed units available on rangeland. Small lambs (14 kg liveweight at 5–6 weeks) are used to supply fattening industries located near Roquefort on the Larzac plateau or in Rouergue. However, present market opportunities are unclear. Unsold lambs are kept, at risk, throughout the summer and sold whenever possible in autumn as heavy lambs. A fair percentage of rams are also sold directly to Muslim families at the Aid-El-Kebir festival. This option works well when the Muslim feast occurs at a favorable time of the year.

In the 'Pre-Alpes' of Provence (Bazin 1984, loc. cit. Bazin and Chassany 1986), the traditional wool and cereal system has been dismantled since the late 19th century when Australian wool production came into the international market. From wool production, the local sheep industry turned to meat production of lambs fed on abandoned terraces once used for cereal cropping or olive growing. Parts of these terraces were also used temporarily for lavender growing to supply the perfume industries of the nearby Cote d'Azur. However, in order to supplement their falling income from lavender and to remain competitive, lamb producers working with small flocks of 50–100 'Merino' ewes had to increase their ovine specialization and the productivity of their flock. This was achieved with higher forage production on irrigated land whenever possible. The average sheep production unit runs approximately 150 'Merinos' or 'Mourerous' or 'Prealpes' ewes on 20 ha (2/3 permanent pasture and 1/3 cereals) and a variable acreage of mountain rangeland.

At present two sheep rearing systems remain in this part of France. They are:

1. *Autumn lambs fattened in sheepfolds*. Lambing takes place in September-October when the well-fed ewes come down from high altitude rangeland. Fed in sheepfold from December onwards, lambs receive an average 500 g/day of hay, and 30–50 kg of cereal grain (barley-maize, wheat-maize) during the 3 months of the fattening process.

2. *Spring lambs fattened on pasture.* Lambing from Merinos ewes takes place in February–March. Lambs are temporarily kept in the sheepfold (March–April), while fed 15 kg of hay and 3–5 kg of cereal grain. Thereafter, they graze permanent pasture and transhume with the ewes to summer ranges in the mountains. They are sold during summer at 30 kg body weight and at 6–7 months old.

   As summer ranges have to be paid for, only shepherds owning such facilities can afford the practice of spring lambing.

A Dual Role for the Sheep Rearing System of the French Mediterranean Regions

Sheep rearing is often the main and dominant activity generating income to the shepherd, but the mutton and lamb production of the French Mediterranean regions is not well placed at the moment to compete with imports on the national market. Lambs and mutton locally produced by small flocks allow only low returns and wages (Paturel and Salmon 1982) in spite of local and national demand. Under these conditions, it is difficult to expect that there will be enough funds to develop fodder crop production, an insurance for a reliable and a sustained income.

In this context, Martinand (1985) differentiates between agricultural units using fodder crops (minimum fodder crop acreage, 4–6 head of sheep/ha) and run by farmers, and pastoral units using only rangeland areas (fodder reserve on rangeland, 0.5–2 head of sheep/ha). A combination of both types often occurs in the south of France.

The first intensive type may satisfy the high fodder requirement of fast growing sheep and lambs, at least in spring, and could expand on a sound and self-supporting economic basis.

The second extensive type can only provide maintenance resources with cheap feed units. It satisfies limited production goals not presently economical and viable. It may, however, also play a second role in maintaining fallows and rangeland in good condition, i.e. clear of bush encroachment, hence not susceptible to bush fire. This type of agricultural activity would then be, to some extent, financially supported by the local or national community. In order to minimize the cost of such support, rangeland improvement may become necessary, even if not an entirely economic proposal.

In any case, better utilization of rangeland is subordinate to intensification and development of the 'minimum fodder crop acreage' which guarantees minimum reliable forage production during difficult periods of the year. It allows the grazier to increase the size of his flock, and hence increase his productivity.

Maintenance is assumed by forage availability on rangeland which is not a limiting factor.

Present Role of Legumes in Farming Systems

Forage legumes such as (lucerne, sainfoin, vetch,) or grain legumes (chickpea, lentil) used to play an important role in various farming systems of the Mediterranean French regions. Their present use is extremely limited for reasons previously detailed. Statistics show their limited extent in the region (Table 3).

Other legumes (soya, forage peas) have recently been introduced in intensive cropping systems of the region.

Perennial lucerne (*Medicago sativa*) of northern origin (Flemish cultivars) is cultivated by farmers on the best soils, mostly on seed contract. Hay production is sold locally to livestock industries or to dehydration units (10%). Lucerne is also occasionally cultivated by vine growers, on small acreages for several years after vines have been uprooted, in order to restore soil fertility. In this case, hay may or may not be harvested. Vetch (*Vicia* sp.) is grown for seeds and sold to other parts of France. Multipurpose sainfoin (*Onobrychis sativa*) is produced in Provence, for hay making, direct grazing and seed production. Sainfoin seed production yields 0.5 t/ha (seeds in husk) and about 150–200 tonnes is sold annually.

Half the acreage and production of forage peas (*Pisum* sp.) is used by factories making concentrates to feed sheep whose milk supplies the milk industries and to fatten lambs in feed lots. Soya bean grown under irrigation in summer is similarly used. Both peas and soya are cropped in rotation with cereal in well-equipped farms where plant production is dominant.

The Potential Place and Role of Some Promising Forage Legumes for the French Mediterranean Zones

The sheep industries of the French Mediterranean regions should increase their contribution to the local and national meat supply in order to reduce lamb and mutton imports. Therefore, they have to mimimize their expenses within the present system of production, and increase their productivity. This could be partially achieved by increasing forage production. This is not

Table 3. Main forage and grain legumes grown in the French Mediterranean regions (1000 ha) compared to France as a whole.

<table>
<tr><th></th><th>Lucerne</th><th>Vetch</th><th>Sainfoin</th><th>Chickpea</th><th>Soya</th><th>Pea</th></tr>
<tr><td>Languedoc-Roussillon</td><td>22.7</td><td>0.1</td><td>0.0</td><td rowspan="3">0.4</td><td>4.2</td><td>4.0</td></tr>
<tr><td>Provence-Cote d'Azur</td><td>25.0</td><td>0.6</td><td>2.0</td><td>1.4</td><td>1.3</td></tr>
<tr><td>Corsica</td><td>2.5</td><td>0.0</td><td>0.0</td><td>0.0</td><td>0.0</td></tr>
<tr><td>France</td><td>500.3</td><td>4.7</td><td>NA</td><td>0.5</td><td>78.5</td><td>426.6</td></tr>
</table>

Source: SCEES (1988).

an easy option due to limited farm machinery, storage facilities, know-how and shortage of capital investment. A gradual improvement strategy relies on quality forage directly available in quantity for direct grazing of cheap feed units, with concomitant measures such as fencing, control of the breeding cycle and higher prolificity of ewes. In the efforts towards increasing forage production, perennial and annual legumes each have a specific place and role.

Perennial Forage Legumes

Lucerne cultivars (*Medicago sativa* cv. Magali, pool 'Provence' etc) for Mediterranean environments are available. However, direct grazing by sheep is limited by severe bloating problems encountered in the south-east of France. Its use could be increased as a mere supply of good quality hay for difficult periods. The potential of rhizomatous and prostrate types of *Medicago falcata* (Spanish 'Mielga' type) is still to be explored.

Sainfoin (*Onobrychis sativa*) is coming back into favor; direct grazing raises no fears of bloating problems. Hay making is possible when the farmer-grazier is equipped with farm machinery. Concomitant seed production, even with low yield, is valuable supplementary income.

Both these crops need arable land and could expand on intensive systems with available farm machinery.

Psoralea bituminosa, locally named 'the sheep fattener' ('Engraisse-mouton') raised real interest for direct reseeding into the Garrigue of Montpellier (Delabarre 1984) and on rangeland. Its low seed yield appears to be a major hindrance to its development however.

Perennial Forage Legumes

A wealth of small annual legumes (annual *Medicago* sp., *Hippocrepis* sp., *Scorpiurus* sp., *Trifolium subterraneum* ssp., *Ornithopus* sp.), occur in the native vegetation. None of them is used up to its potential in spite of its wide acceptance in other parts of the world (Puckridge and French 1983).

Perennial forage legumes are cropped in the south east of France, annual forage legumes are still to be discovered, though well known by shepherds and grazed when available.

Rangeland and Fallows

Several attempts have been made to draw the attention of sheep producers to subclovers, either for regions with acid soils in Corsica or Provence-Cote d'Azur (Etienne 1977 and Masson and Gintzburger 1987). Similarly, annual medics have been recommended for other soils by Dauro and Gintzburger (1987), and Gintzburger and Prosperi (1987). Most efforts aimed initially at improvement of rangeland and fallows, the assumption being that animals fed on nitrogen-rich forage such as legumes increase their consumption of ligneous vegetation (Genin 1985; Goby et al. 1987).

Increasing availability of legumes can be achieved either by stimulating annual legumes of the native flora with phosphate fertilizer or by reseeding with available cultivars.

Dauro and Gintzburger (1987) reported a significant increase in native annual legumes' phytomass with one single broadcast of simple superphosphate applied in autumn. During the following spring, annual legume phytomass (mostly *Medicago rigidula*) increased from 188 (under zero P_2O_5) to 812 and 1240 kg D.M./ha with applications of 80 and 400 kg P_2O_5/ha, respectively. No response from grasses or other plants was observed. The relative specific contribution of annual legumes moved from 12% to nearly 40% of phytomass. The contribution of grasses dropped from 68% to less than 50%. Yield components of *M. rigidula*, and especially pods and seeds production, increased significantly, including a higher contribution of these legumes during the following years without further application of phosphate fertilizer. By late May, at early pod maturity of annual legumes, their total nitrogen content (Kjeldhal technique) was about 3.25% while grasses reached only 1.75% of dry matter weight. Estimated cost of one single application of 80 kg P_2O_5/ha, the effect of which would last at least 3–4 years, is about 446 FF/ha (90 USD/ha).

Such a simple and cheap technique may easily and quickly contribute to improved forage quality and quantity for direct grazing on native pasture containing some annual legumes. Sheep rearing systems using rangeland could consider this approach favorably.

In Vineyards

Reseeding is advocated by Masson and Gintzburger (1987) not only for rangeland improvement in Roussillon, but also in sheep rearing systems grazing vineyards during autumn to early spring in the same province. Traditional grazing of vineyards could benefit from interseeding vines with either subclovers or appropriate medics. The winter development of Mediterranean annual legumes will not interfere with vine growth and development. Reciprocal benefits are expected for both the vine grower and the shepherds. Benefits for growers include: weed control by winter grazing, improved soil fertility from legumes and through sheep manure, a mulching effect during the summer, and firmer soil surfaces during grape harvest. The shepherd would obtain high quality feed during autumn and winter. Phytomass measured (Goby et al. 1985) with Subclovers cv. Nungarin, Daliak, and Clare reached 750 to 950 kg DM/ha by mid-January in first year regeneration, compared to 300–800 kg of weeds (50% *Lolium rigidum* dry weight). However, seeding annual legumes is not yet accepted by vine growers for fear of future problems of weed control and the apparent complexity of the new system, including official grazing in between vines.

Towards Integration of Annual Legumes Into a Forage System

Proper use of annual legumes as forage in sheep rearing systems of the south of France has still to be demonstrated.

Within this environment, annual pasture legumes have to be properly grazed during autumn, winter and spring. In the first two seasons, expected production ranges from 0 to 40 kg DM/day/ha. In spring, it reaches 60 to 250 kg DM/day/ha.

In spite of their low winter production, annual legumes offer a forage of high quality (20–25% crude protein).

In a preliminary experiment, Rochon et al. (1987) used a native pasture containing about 30–40% of native medics (*M. polymorpha*, *M. arabica*) to feed ewes and fatten their lambs without supplementary feeding. The experiment was conducted for three months, from November 1986 to January 1987. During this period, average daily growth of pasture was 26 kg DM/ha, and stocking rate was four ewes with lamb/ha. The lambs fed on native medic pasture averaged a daily gain of 270 g and their body weight averaged 27 kg at slaughtering when they were 80–100 days old. Thus they compared well with the local sheep-fold fattening system of Perpignan. At the same time, ewes and lambs of the traditional sheep rearing system of the plain of Roussillon are kept exclusively in sheep-pens and are fed 25 to 60 kg concentrates/ewe with lamb.

These results mean that native pasture of annual medics freely available to some shepherds in the plain of Roussillon could favourably replace concentrates bought every winter.

Unfortunately, such a lamb production scheme requires large areas of annual legumes inconsistent with the average size of a productive flock in the region. Therefore, it is suggested that in winter, fodder production from annual legumes pastures should be managed in the same way as concentrates. Early grazing must be allocated to lactating ewes during the first month after delivery, and then for fattening weaned lambs. For example, in February 1988, weaned lambs 100 days old were fed for 48 days on a lean native medic pasture (530 kg DM/ha average phytomass). They achieved a growth rate of 181 g/day when supplemented with only 325 g of barley grain/day/lamb. Stocking rate was 20 lambs/ha. In spite of the low phytomass available, quality was high enough to cover nearly 80% of energy requirements.

Later in the season from mid-March to early June, annual legumes pastures still present a higher growth rate (80 to 260 kg DM/ha/day) than native pasture with a dominant tall fescue and couch grass (60 to 120 kg DM/ha/day. Crude protein during the same period is on average slightly higher for legumes pastures (11.4%) than for grasses (9.8%). Ewes and lambs feeding on annual legumes pastures in spring will assist in releasing other pastures for hay stored for animals with low feed requirement in early summer, or early autumn after transhumance.

However, with such a grazing system, management techniques for annual legumes with appropriate hardseededness have still to allow a reasonable seed set and ensure a reliable pasture regeneration year after year.

Conclusions

Sheep rearing systems for meat production in the French Mediterranean regions are in a difficult position. If they persist with traditional shepherding systems harvesting cheap sheep units from rangelands, they will slowly disappear because of their low productivity compared to other sheep rearing systems in France, or other countries inside or outside the EEC. Increasing their productivity is a must for survival. Land tenure systems and competition for land from hunting societies may cause problems for their development. However, forage resources available in quantity and quality mainly during autumn and winter is the main bottleneck. Such availability would allow production of fast growing lambs requiring little supplementation with concentrates, thus reducing expenses and increasing productivity of the flock.

Perennial legumes, though useful for hay making, are bound to cropping systems and the best agricultural land.

Introduction of annual forage legumes into forage systems, either on agricultural land or on the waste and rangelands of the French Mediterranean regions, show some promise still to be explored.

References

Barello, A., Laberche, J.C., Masson, Ph., Pavageau, J. and Rochon, J.J.1984. Les exploitations agricoles de l'Aspre et du Fenouillede et leur intervention sur le milieu. IUT - Agronomie, Perpignan, France. 65 p.

Bazin, G., and Chassany, J.P. 1986. Quelles perspectives pour l'elevage ovin dans les montagnes seches? E.S.R., Grignon et Montpellier, France. 42 p.

Boutonnet, J.P. 1981. L'elevage ovin de la zone mediterranenne francaise: Influence des rapport au foncier et des marches – Evolution historique et situation actuelle. E.S.R. INRA, Montpellier, France. 46 p.

Cavalier, J.B. 1983. Dynamique des systemes de production a base delevage en Garrigue et recherches sur les ameliorations pastorales. E.S.R. – INRA, Montpellier, France. 114p + annex.

C.N.R.S. 1985. Les Garrigue du Montpellierais: de la marginalisation a la periurbanisation (Etude ecologique, sociologique et economique d'une zone mediterraneenne mediane), Programme P.I.R.E.N.-C.N.R.S., 186 p.

Dauro, D. and Gintzburger, G. 1987. Effet de la fertilisation phosphatee sur parcours naturel contenant des luzernes annuelles 9en region mediterraneenne francaise et froide. Pages 23–28 *in*: FAO-European cooperative network on pasture and fodder crop production. Bulletin 5. Montpellier, France.

Delabarre, Yvette, 1984. Contribution a l'etude de *Psoralea bituminosa* L. en vue d'une amelioration fouragere des parcours de Garrigue. E.N.S.A.-U.S.T.L., Montpellier, France. 69 p.

Etienne, M. 1977. Bases phyto-ecologiques du developpement des ressources pastorales en Corse. These U.S.T.L. Montpellier, France. 210 p.
Genin, D. 1985. L'animal debrousailleur en foret mediterraneenne: Essai de realisation technique dans la suberaie des Alberes. C.E.P.E.-C.N.R.S., Montpellier, France. 69 p.
Gintzburger, G. and Prosperi, J.M. 1987. D'autres luzernes annuelles. Bull. F.N.A.M.S. Semences, 101: 23–26.
Goby, J.P., Laberche, J.C., Masson, Ph. and Rochon, J.J. 1985, Analyse de quelques systemes d'elevage de la zone mediterraneene des Pyrennees Orientales, IUT – Agronomie, Perpignan, France. 48 p.
Goby, J.P., Masson, Ph. and Seigneurbieux, C. 1987. Perspective d'entretien des suberaies debrousaillees et sursemees en Trefle souterrain par des troupeaux caprins. Pages 6–10 *in* FAO-European cooperative network on pasture and fodder crop production. Bulletin 5. Montpellier, France.
Mansat, P. 1980. Project de developpement d'un laboratoire d'Amelioration des Plantes Fourrageres pour la zone mediterranneene. INRA, Montpellier, France. 24 p.
Masson, Ph. and Gintzburger G. 1987. Le trefle souterrain (*Trifolium subterraneum*): Essai preliminaires et perspectives d'utilisation dans une zone mediterraneene francaise: le Roussillon. Fourrages, 110: 183–204.
Martinand, P. 1985. Pratiques pastorales et strategies d'exploitations dans les moyennes montagnes mediterraneennes. Bull. Tech. Inf. 'Special montagne', 399–401.
Paturel, Michele and Salmon, C. 1982. L'economie d'un troupeau ovin en Garrigue au cours de l'annee 1980–81. Economie Meridionale 118: 27–29.
Prudhomme, P. 1986. Dynamique de l'interface Elevage-Espace des Garrigue de la region d'Aumelas. LECSA-INRA, Montpellier, France. 105 p.
Puckridge, D.W. and French R.J. 1983. The annual legume pasture in the Cereal-ley farming system of Southern Australia: a review. Agricultural Ecosystems and the Environment 9: 229–267.
Rochon, J.J., Goby, J.P. and Gintzburger G. 1987. Essai de production d'agneaux sur prairies a luzernes annuelles indigenes en plaine du Roussillon. Pages 18–22 *in* FAO-European cooperative network on pasture and fodder crop production. Bulletin 5. Montpellier, France.
Rouche, J. 1987. Bilan et perspectives d'un elevage ovin en Garrigue. Agrimed-Cemagref, Montpellier, France. 102 p.
SCEES 1983. Graph Agri Regions 1983, Annuaires de Graphiques agricoles, Ministere de l'Agriculture, Paris, France. 255 p.
SCEES 1988. Annuaire de Statistique Agricole, Resultats provisoires 1987. Ministere de l'Agriculture, Paris, France. 227 p.

The Role of Forage Legumes in Rotation with Cereals in Mediterranean Areas

M. J. JONES

International Center for Agricultural Research in the Dry Areas (ICARDA), P.O. Box 5466, Aleppo, Syria

Abstract. In the context of the two broad types of farming systems in WANA, barley-based in the drier areas and wheat-based in the wetter areas, this paper defines what is meant by forage legumes, describes the three main ways in which they are utilized as animal feed, and explains how they fit into the crop sequence. It concludes by outlining future research issues, first to identify the most suitable legumes, and then to find the most effective means of persuading the farmer to adopt them.

Introduction

The objective of this workshop is to review the role of pasture, forage and food legumes in the farming systems of Mediterranean areas, and that of this paper to focus specifically on the forages; but, first, what farming systems are there? Leaving aside irrigated agriculture and high-altitude agriculture, two broad types may be identified across North Africa and West Asia: 'wheat-based' in wetter areas, 'barley-based' in drier areas. With some local differences due to elevation, soil depth, and soil type, the transition between them lies around the 300 mm isohyet (Fig. 1).

Although the most obvious distinction between these systems lies in the predominant cereal crop, two other differences are, arguably, more fundamental. The first relates to production goals. In the wheat-based system, the major agricultural aim is to grow crops. Some animals are kept, for their produce and for sale; but the main goal of most farmers is the production of crops, for subsistence and for sale. In contrast, in barley-based systems, the main goal is animal production. Where conditions permit, wheat is grown for subsistence; but with decreasing rainfall and rainfall reliability, the farmer places increasing reliance on animals, mainly sheep and goats. In times of drought and scarcity, animals (unlike crops) can be moved to alternative feed and water sources, or some can be sold to pay for feed for the others.

The second difference concerns secondary crops. In wheat-based systems, these are relatively numerous; but in many drier areas, barley and (to a

A.E. Osman et al. (eds.), The Role of Legumes in the Farming Systems of the Mediterranean Areas, 195–204.

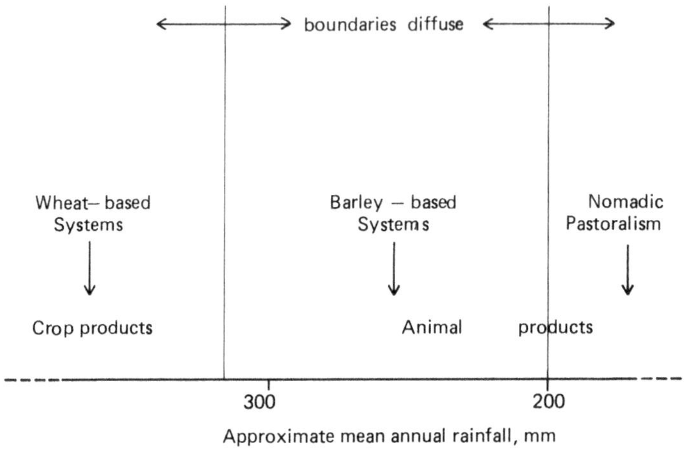

Fig. 1. Sequence of rainfed farming systems across the rainfall gradient in Syria.

much lesser extent) wheat are the only crops ever grown (Fig. 2). In Syria, lentils, the most drought-resistant of the grain legumes, are rarely seen below the 300 mm isohyet (Tully 1984): similarly in Morocco (Jouve 1978). In a survey in Idleb Province, Syria, Nordblom (1987) identified 24 different crop rotations practised in areas of relatively high mean annual rainfall (approx 340–580 mm) but only 10 in drier areas (250–330 mm). Three-course rotations of cereals, grain legumes (lentils, chickpeas) and summer crops (melons, sesame) predominated in the wetter areas; two-course rota-

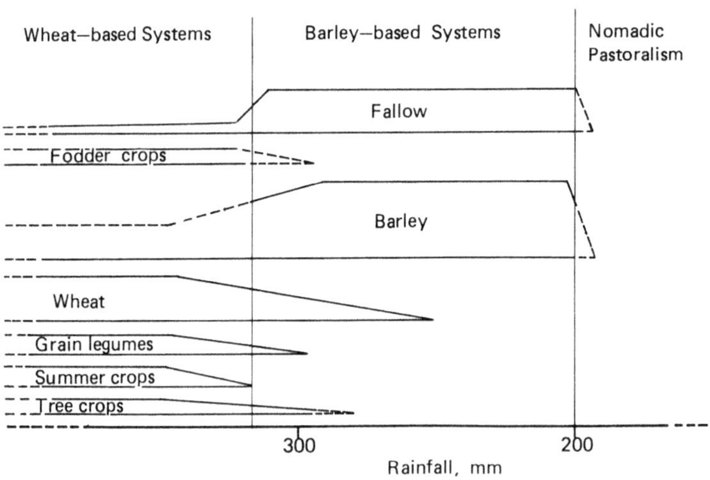

Fig. 2. Semi-quantitative representation of crop areas in wheat- and barley-based farming systems in Syria.

tions (mainly cereal-fallow) in drier areas. Forage legumes (lathyrus and vetch), though part of the rotation in nearly 9% of the fields surveyed, were limited to higher rainfall areas. It is the thesis of ICARDA research and of this paper that forage legumes have an important role to play in lower rainfall areas also; and it is on these areas that we shall concentrate here.

Few farming systems, whether wheat-based or barley-based, are static. Rather, they tend to evolve locally in different ways, according to a variety of local pressures – increasing human populations and land shortages, increasing mechanization, and changing market forces. For example, the Breda/Bueda area of Aleppo Province, Syria, with mean annual rainfall around 280 mm, now supports a typical barley-based farming system. A survey in 1982/3 showed 45% of the arable land to be under barley, 19.5% under wheat and 32% in fallow (with 1.5% summer crops and 2% legumes); but as recently as the 1950s wheat had comprised 90% of the cereal area, barley only 10% (Jaubert and Oglah 1985). The change arose directly from an increased need for feed production. Previously, sheep and goats derived much more of their feed from natural grazing, locally on hillslopes and fallow land and, at a distance, in the steppe. Increasing populations and the availability of tractors have caused larger areas to be ploughed, with a consequent reduction in all types of grazing land. Flocks have become more sedentary and, hence, more dependent on local barley crops: grain and straw in the winter, but also green grazing in the spring and stubble grazing in the summer. Barley has, therefore, largely displaced wheat, and farmers in turn have become more dependent on the sale of livestock products for their livelihood.

Concurrently, the relative area of fallow has declined, as farmers increasingly sow barley on the same land year after year; and there has been an almost total loss of forage legumes (mainly lathyrus) from the system. In the 1950s, the area of forage legumes ranged between 10–15% of the rainfed cultivated area, while at present it is only 2%, i.e. a loss of 8–13%. This loss is attributed to increased labour costs and, ironically, to the supply by the General Organization of Feed of agricultural by-products as supplementary feed (Jaubert and Oglah 1985). These were intended originally just as a buffer against drought. Such changes are worrying. Barley monocropping is unlikely to provide stable and sustained yields in the long term; and too great an evolution towards a feedlot farming system, heavily dependent on fluctuating external feed sources, appears equally dangerous. Far better to seek to sustain long-term animal production by optimizing local feed production from barley and legumes.

We believe that this area epitomises the situation of dry-zone, rainfed farming through much of the Mediterranean region: its major dependence on animals, its decline of natural grazing resources, and its increased need for feed production. In such areas, there is an urgent requirement for forage legumes able to compete, in terms of quantity and quality, with barley and able to comprise stable rotations with barley.

Definition of Forage Legumes

But what are the forage legumes? Many species of *Vicia* (vetch) and lathyrus are native to Mediterranean areas and provide excellent feed, and some of them have long been cultivated. *Pisum* species are also said to make excellent feed, although Syrian sheep have not so far been persuaded to eat them (Thomson 1988). Sheep do eat pisum in Morocco, but there, the peas are also valued as human food. This reminds me that in Europe we have two sorts of faba beans, field bean for stock feed and broad beans for human consumption – both, of course, *Vicia faba* and related to *Vicia narbonensis*, a species that is doing very well in ICARDA forage trials in dry areas. It reminds me also, that not only is lentil straw a highly valued fodder, but in Syria lentil grain is used by some farmers to fatten lambs, and poor lentil crops may be grazed at maturity *in situ* (Nordblom 1987). It seems that the only clear distinction between forage and food legumes lies in the mode of their utilization in any particular situation. There is the potential in some situations for there to be competition between humans and animals for legume grain, but that will be resolved ultimately by economics.

We may also question whether there is any fundamental difference between forage and pasture legumes. El Moneim (1987) has recently identified a variety of vetch (*Vicia sativa*, spp. *amphicarpa*), naturally occurring in Turkey and Syria, that produces both aerial and subterranean seeds. More recently still, he has acquired a subterranean lathyrus (El Moneim, personal communication). Undoubtedly, a lot of research and testing lies ahead, but we may anticipate the eventual possibility of using vetch and lathyrus, like medicago, as self-regenerating annual legumes in pastures or in rotation with cereals.

It seems to me as an agronomist (and therefore an outsider in specialist legume deliberations), that new, exciting opportunities for legumes in Mediterranean farming systems are in view. However, I think they will more likely be realised if labels like food and fodder, and pasture and forage, are applied less rigidly to particular species and more simply to describe actual utilization.

Utilization of Forage Legumes

Utilization has another, narrower connotation. Even where a legume is grown unambiguously to feed animals, there is still the question of how and when it is best used. In the Mediterranean context, at least, three main modes may be distinguished:

1. *Green grazing*, e.g. to fatten lambs or increase milk production in the spring. There are advantages here in the lack of harvest costs (beyond those of shepherding) and the reduction of pressure on natural pastures at a time when good seed set is essential to the survival of native annual species (Thomson 1987), but lack of grain production for seed is a disadvantage.

2. *Hay-making* conserves good quality feed for the winter, but labour and equipment costs may be high and yields low (Thomson and Oglah 1988), and again no seed is produced. Rihawi et al. (1987) have commented that it is difficult to make hay with forage legumes because of their thick stems, slow drying, and tendency to lose leafy material (although grazing the aftermath can reduce such losses); and for both modes 1 and 2 there are arguments for using a legume/cereal mixture instead of a pure legume (Osman and Nersoyan 1986).
3. *Mature harvesting* also conserves high quality feed for the winter (and seed for subsequent planting) but requires a heavy labour input at a peak period. Farmers in the Breda/Bueda area cited as reasons for the decline in lathyrus cropping, the need to harvest during a very short period before maturity to avoid pod shattering, with hand-pulling to collect all the straw (Jaubert and Oglah 1985). From a survey near Al Bab (Syria), Tully (1984) concluded that the harvest labour problem for legumes (lathyrus, lentils and vetch) is primarily one of cost relative to crop value; and at Breda farmers prefer to limit forage legumes to less than two hectares, which can be harvested by members of the family (Thomson and Oglah 1988). Current research on mechanized lentil harvesting seems likely to have eventual application to other legumes grown to maturity.

There is potentially a fourth mode, sometimes practised by default, which is to graze the mature crop as standing hay; but losses due to desiccation and shattering are usually high, and seed for subsequent planting is not collected.

No single one of these modes will prove generally superior. Different forms of utilization will be appropriate in different circumstances. It is here that feedback from the farmer, via on-farm trials, is particularly important, as the work of Thomson and colleagues at ICARDA has shown (e.g. Thomson and Oglah 1987, 1988). The farmer needs to see the options and make choices according to his production aims, but those choices have relevance for the future course of research. Currently, in Syria, although the high cost of seed is a major consideration, high prices for fattened lambs and milk products may still favour green grazing.

Mode of utilization may also have implications for the crop rotation as a whole – for example, on time of tillage; on the amount of water left in the soil profile for the following crop; and, in the longer term, on the weed flora. Effects on the amount of fixed nitrogen left in the soil would also be expected.

Fodder Legumes in Crop Sequences

Legumes are universally regarded as beneficial to crop rotations, on the general assumption that they improve soil fertility by adding nitrogen; but the magnitude of this addition is rarely measured unambiguously. Two recent comparisons of N-uptake by barley crops following legumes or barley suggest a wide range of differences between environments and legume species on the enhancement of subsequent nitrogen availability (Table 1); but in Cyprus, under good rainfall and with phosphate fertilization, a preceding vetch crop increased the N-uptake of barley by 35–50 kg/ha (i.e. 100–200%).

However, such comparisons may inflate the apparent legume effect. The performance of barley following barley is often depressed by adverse biological factors, e.g. pest build-up and growth inhibition by phytotoxins (Krause et al. 1988); and soil nitrogen availability may be positively depressed by cereal crop residues. In another trial at Breda, barley following vetch took up less nitrogen than barley following fallow (Table 2). Part of the explanation lies in the tendency of fallows to accumulate available nitrogen (and possibly other nutrients too); but this result is a reminder that, depending on the rotation previously practised, the legume in legume-cereal rotations may be seen from different perspectives: as replacing fallow in a cereal-fallow rotation or as replacing cereal, in alternate years, in a continuous cereal rotation.

Table 1. Effect of previous crop on the nitrogen uptake of a barley test-crop.

Data source	Previous crop	N-uptake, kg/ha		Incease relative to barley-barley, kg/ha	
		1983/84	1984/85	1983/84	1984/85
1	Barley	28.1	24.9		
	Lathyrus ochrus	45.7	54.6	17.6	29.7
	Vicia dasycarpa		83.6		58.7
	Vicia ervilia	44.6	71.0	16.5	46.1
	Medicago truncatula	42.1	63.8	14.0	38.9
	Chickpeas		45.5		20.6
2	Barley	32.3			
	Pisum sativum	36.8		4.5	
	Vicia sativa	38.1		5.8	

Sources: 1 Papastylianou (1987). Site: Dromalaxia, Cyprus. Test-crop received 20 kg P/ha but no N; 2 Keatinge et al. (1988). Data means of three sites, Breda, Tel Hadya and Jindiress, Syria. Test-crop received 39 kg P/ha but no N.

Table 2. Total dry matter production (t/ha) and nitrogen uptake (kg N/ha) of unfertilized barley following fallow or vetch (*Vicia sativa*) at Breda, Syria.

Previous crop	1982/83		1983/84	
	DM	N	DM	N
Unfertilized barley	1370	9.3	880	8.3
Fallow	2030	20.8	1840	25.2
Unfertilized vetch	1820	17.3	1470	20.0
P-fertilized vetch	2840	17.3	2370	24.8
Rainfall, Nov–April, mm	231		196	

Source: Rashed (1986).

Generally, it is better to consider the whole rotation over a period of years. From Cyprus, Papastylianou (1988) reported the cumulative crop output of nitrogen over five years, in two rotations:

Barley for hay – barley for grain	205 kg N/ha
Vetch – barley for grain	285 kg N/ha
(values read from figure)	

The vetch-barley rotation received no nitrogen fertilizer, the barley-barley rotation 60 kg N/ha in each phase. Although part of the success of the barley-vetch rotation may have derived simply from the alternation of crops, which minimized pest and phytotoxic effects, there can be little doubt which rotation had the better nitrogen economy.

Dry matter data from Syria tell a similar story (Table 3). Mean productivity from a barley-vetch rotation exceeded that from a barley-barley rotation, by 39% without fertilizer, and by 15% with fertilizer (with the barley-barley but not the vetch-barley receiving N-fertilizer annually). And calculations based on crop contents of crude protein and metabolizable energy, indicate theoretical sheep-carrying capacities 46–57% higher in barley-vetch compared with barley-fallow rotations (Jones 1988).

Future Research Issues

At its simplest, the problem is first to identify the right legumes and then to persuade the farmer to use them.

The right legumes must:

a) tolerate/resist the constraints of the environment, e.g. often low and unreliable rainfall; low winter temperatures (in many areas); the local spectrum of pests (including *Orobanche*, *Sitona* weevil, leaf-eating birds);

Table 3. Six-year mean total dry matter yields from three rotations under different fertilizer regimes at Breda, Syria.

Fertilizer regime		Rotation		
		Vetch-barley	Fallow-barley	Barley-barley
Zero	DM, t/ha	1.66	1.12	1.19
	sd (+/−)	0.326	0.389	0.188
	cv (%)	19.6	34.8	15.8
Biannual, 20N : 26P to barley only	DM, t/ha	2.87	2.34	
	sd (+/−)	0.802	0.443	
	cv (%)	27.9	19.0	
Annual, 20N : 26P to barley, 26P to vetch	DM, t/ha	3.43		2.98
	sd (+/−)	1.050		1.035
	cv (%)	30.7		34.8

Source: Jones (1988).

b) be palatable to animals;
c) possess characteristics that match the likely utilization:
   - if for green grazing, then vigorous cold-season growth;
   - if for hay, then appropriate physical characteristics for hay-making (thin stems, low leaf loss) or an ability to grow well in mixtures with cereals;
   - if for mature harvest, then a high harvest index and good resistance to shattering.

   (Since many farmers are likely to use the same legume in different ways at different times, a high rating in all these characteristics will usually be desirable.)

However, the right legumes must also meet the farmer's needs. They must demonstrably outproduce either

- barley in a continuous barley system (or, sometimes, a wheat-barley system), or
- fallow in a fallow-barley system,

without increased net cost of money and labour. And they must not diminish overall feed security. Barley, even in continuous rotation, is a versatile crop, able to provide green grazing and stubble grazing as well as winter grain and straw. A forage legume, replacing it in alternate years (or on alternate fields), may produce more or better feed (and, in the long term, greater yield stability); but to the farmer it may appear to reduce his feed options and leave gaps in the annual feed cycle.

Where the legume is replacing what would previously have been fallow, there should be fewer negative features, since – in West Asia, at least – fallows contribute little forage to the feed cycle. In North Africa, where weedy fallows are more productive, a forage legume might well be uncompetitive as a provider of green grazing but still of high potential value as a source of conserved feed for the winter.

Scientists should try to anticipate these factors and, hence, farmers' reactions to new crops, during selection and development; but ultimately there is no substitute for collaborative on-farm testing with farmers. Such testing is particularly important with forage legumes for the following reasons:

a) crop value needs to be assessed not just as t/ha yield but in terms of the feed value (quantity, quality and time of availability) to the farmers' animals; and
b) the introduction of a new crop is a big step, even in an evolving farming system. Asking a farmer to try a new variety of a known crop or a fertilizer or herbicide does not require him to change his system; but a new crop does. Such a change will likely involve new requirements in equipment and labour inputs and innovations in his seasonal patterns of feed utilization, purchase and (in some cases) sale. Such changes need to be monitored, constraints to adoption investigated, and the information fed back into the research process.

References

El Moneim, A.A. 1988. Preliminary evaluation of subterranean vetch (*Vicia sativa*, spp. *amphicarpa*). Pages 179–180 *in* Annual Report 1987, Pasture, Forage and Livestock Program, ICARDA, Aleppo, Syria.

Jaubert, R. and Oglah, M. 1985. Farming system management in Bueda/Breda subarea, 1983/1984. Research Report No. 13, Farming Systems Program, ICARDA, Aleppo, Syria.

Jones, M.J. 1988. Barley rotation trials at Tel Hadya and Breda stations: a summary of biological yield data, 1981–87. ICARDA report (in preparation).

Jouve, P. 1978. Analyse des modes de conduite des cereales et voies d'amelioration des rendements en zones semi-aride et aride du Maroc occidental. Departement d'Agronomie, Institut Agronomique et Veterinaire Hassan II, Rabat, Morocco.

Keatinge, J.D.H., Chapanian, N. and Saxena, M.C. 1988. Effect of improved management of legumes in a legume-cereal rotation on field estimates of crop nitrogen uptake and symbiotic nitrogen fixation in northern Syria. Journal of Agricultural Science, Cambridge 110: 651–659.

Krause, S., Weltzien, H., Mamluk, O. and Cocks, P.S., 1988. Yield decline in continuous cereal systems, Pages 218–229 *in* Annual Report 1987, Pasture, Forage and Livestock Program, ICARDA, Aleppo, Syria.

Nordblom, T.L. 1987. Farming practices in southern Idleb Province, Syria: 1985 survey results. Report, ICARDA - 107 En, ICARDA, Aleppo, Syria.

Osman, A.E. and Nersoyan, N. 1986. Effect of the proportion of species on the yield and quality of forage mixtures, and on the yield of barley in the following year. Experimental Agriculture 22: 345–351.

Papastylianou, I. 1987. Effect of preceding legume or cereal on barley grain and nitrogen yield. Journal of Agricultural Science, Cambridge 108: 623–626.

Papastylianou, I. 1988. The role of legumes in agricultural production in Cyprus. Pages 55–63 *in* Nitrogen Fixation by Legumes in Mediterranean Agriculture. Proceedings of a Workshop, Aleppo, 1986 (Beck, D.P. and Materon, L.A., eds). Martinus Nijhoff, The Hague, Netherlands.

Rashed, E. 1986. The effects of fertilization and crop rotation on rainfed barley development, growth and yields in a semi-arid Mediterranean climate. Unpublished PhD Thesis, McGill University, Montreal, Canada.

Rihawi, S., Capper, B.S., Osman, A.E. and Thomson, E.F. 1987. Effects of crop maturity, weather conditions and cutting height on yield, harvesting losses and nutritive value of cereal-legume mixtures grown for hay production. Experimental Agriculture 23: 451–459.

Thomson, E.F. 1987. Feeding systems and sheep husbandry in the barley belt of Syria. Report, ICARDA-106 En, ICARDA, Aleppo, Syria.

Thomson, E.F. 1988. Screening of forage peas for palatability. Pages 85–87 *in* Annual Report 1987, Pasture, Forage and Livestock Program, ICARDA, Aleppo, Syria.

Thomson, E.F. and Oglah, M. 1987. On-farm forage and livestock experiments at Breda, Pages 62–71 *in* Annual Report 1986, Pasture, Forage and Livestock Program, ICARDA, Aleppo, Syria.

Thomson, E.F. and Oglah, M. 1988. On-farm experiments with forage legumes – six years results from Breda. Pages 74–82 *in* Annual Report, 1987, Pasture, Forage and Livestock Program, ICARDA, Aleppo, Syria.

Tully, D. 1984. Land use and farmers' strategies in Al Bab: the feasibility of forage legumes in place of fallow. Research Report No. 12, Farming Systems Program, ICARDA, Aleppo, Syria.

Discussion

Durutan:

In the barley-vetch rotation trials at Breda did you measure the soil moisture and relate this to the nitrogen response?

Jones:
Yes, the soil moisture was measured in one of the trials and the results are being prepared for publication. They have not yet been related to the nitrogen response.

Capper:
It may be possible to manipulate feeding and breeding management of Awassi sheep so that lambing coincides with time of maximum forage availability. This is done in other parts of the Mediterranean with different breeds of sheep.

Mawlawi:
Should legumes be grown as a pure stand or as a mixture with cereal?

Jones:
It is easier to handle and harvest a mixture, but there may be a build-up of pests and diseases without the 'break crop'. My results show that there was greater productivity with a barley/vetch mixture but the barley following in the rotation gave less. There are not enough results to be sure yet. Farmers should be given the option after demonstrations.

Tawil:
What is the lowest rainfall limit for your work in the drier areas? What is the seed rate for barley? What is the expected yield of forage crops?

Jones:
Definitely not below 200 mm; some people would put it higher, at 230–240 mm. The seed rate for barley is 90 kg/ha. The expected yield for vetch is very dependent on rainfall, of course, but on average hay would be about 2 t/ha and the mature crop a bit more.

Haddad:
When we consider the effect of legumes on the improvement of soil nitrogen we sometimes forget that in dry years, which happen frequently in our area, the legume will probably make no contribution. Also, the experiments described deal only with two-year rotations: in many areas of the region a three-year rotation is often practised, where summer vegetables or tobacco are grown in the third year. I suggest further studies should include three-year rotations, especially with grain legumes.

The Role of Legumes in Improving Marginal Lands

A.E. OSMAN,[1] M. PAGNOTTA,[1] L. RUSSI,[1] P.S. COCKS[1] and M. FALCINELLI[2]

[1]*ICARDA, Aleppo, Syria;* [2]*Institute of Plant Breeding, University of Perugia, Italy*

Abstract. The term marginal land refers to non-arable land within and adjacent to the cropping zone of the Mediterranean basin. This land constitutes a sizable portion of the total land resource available. It is used primarily for livestock grazing but because its productivity is low, animals using it need supplementary feeding from various sources.

Improving the productivity of these lands should reduce the need for supplementation and increase their carrying capacity. Marginal lands were found to be rich in legume species which can be used as the basis for improving productivity. One possibility which is discussed here is to apply phosphate fertilizer, and some of the results obtained from an ongoing long-term experiment established on 80 ha of marginal land in north Syria are presented. On the fertilized plots the legume component, including seed production, increased up to four times, and sheep gained weight faster requiring less supplementary feed.

Introduction

Marginal lands in this paper refer to the non-arable lands which exist within or adjacent to the cropping zone in the Mediterranean basin. Such lands constitute a sizeable proportion of the total land available. In Syria and Lebanon for example they constitute between 30 and 60% of the total land, while in Italy the total area of such land is estimated to be two million hectares. The land is often steep, usually stony and the soils are often shallow. It is intensively grazed and in many cases there is severe soil erosion.

In West Asia and North Africa (WANA), these lands are used mainly for the grazing of sheep and goats. However, because productivity is low the animals need supplementary feeding with barley grain, grazing of green barley or sown forages. Improving productivity should reduce the need for supplementation and increase carrying capacity. In many countries of southern Europe (e.g. Portugal, Spain, France, Greece and Italy) marginal lands represent resources of both ecological and economic value and the research work supported by the EEC and various national programs aims at improv-

A.E. Osman et al. (eds.), The Role of Legumes in the Farming Systems of the Mediterranean Areas, 205–216.

ing farm income as well as protecting the land from soil erosion and fire damage (Veronesi 1987).

Improvement of marginal lands is difficult because they are not arable, often due to their steep terrain, stony surface and shallow soils. Nevertheless, we believe three possibilities exist for improving productivity. The first, and possibly only method applicable to low rainfall areas (<300 mm), is to change grazing management so that grazing pressure is eased during the critical period of seed set. The second method is to change the botanical composition by sowing improved pasture species. The third possibility is to apply suitable fertilizer(s). Both the second and third approaches require favorable climatic conditions and proper grazing management. All three approaches aim at achieving an adequate vegetation cover and suitable plant species.

Presence of legumes is of special importance due to their role in improving the quality of forage as well as restoring soil nitrogen status. In this paper the potential role of legumes in improving marginal land is discussed with special emphasis on the use of phosphate fertilizers.

Soil and Vegetation of the Grazing Land

As in the arable lands, soils of the grazing lands of WANA and south Europe are diverse, but calcareous soils predominate. They vary in texture, depth, slope and stone cover although shallow soils are the most common. Calcium carbonate has a pronounced influence on the chemistry of these soils and deficiencies in major plant nutrients such as nitrogen and phosphorus are widespread. Unlike arable lands, the organic matter content is often high (Cocks et al. 1988): in a survey in north Syria the organic matter content of marginal land was around 4%, compared with 1% on arable land.

The vegetation is predominantly annual, although perennial grasses and legumes species are also important in areas of high rainfall. Legumes are widespread, especially *Medicago* (medics) and *Trifolium* (clovers). Table 1 shows some of the common species on marginal land in Syria and southern Europe, where nineteen clover and fourteen medic species are listed with 31 other legumes, some of them perennials. Of course, many more species are present.

Size of the plant population determines pasture productivity (Abd El Moneim and Cocks 1986) and in the Mediterranean-type pastures of southern Australia, Puckridge and French (1983) have estimated that about 1000 legume seedlings/m^2 are necessary to achieve maximum production. Such populations are rare (though not unknown) in the Mediterranean basin, with seed populations in the marginal lands of northern Syria averaging 8 g/m^2. In a detailed survey of one site in north Syria 246 legume seedlings/m^2 were measured (Ehrmann, T.A.M. and Cocks, P.S., personal communication). As will be shown later, populations of this level are adequate for pasture improvement under 350 mm of rainfall without the need for artificial seeding.

Table 1. List of common legume species on marginal land of northern Syria and southern Europe.

Astragalus triradiata	Ornithopus compressus
A. hamosus	Pisum sativum
A. suberosus (perenn.)	Scorpiurus muricatus
A. asterias	Trifolium campestre
Coronilla scorpioides	T. tomentosum
Hedysarum coronarium (perenn.)	T. stellatum
Hippocrepis unsiliquosa	T. pilulare
Hymenocarpus circinnatus	T. pauciflorum
Lathyrus sativus	T. scabrum
L. aphaca	T. haussknechtii
L. cicera	T. angustifolium
L. inconspicuus	T. cherleri
L. annuus	T. spomosum
Lotus corniculatus (perenn.)	T. argutum
Medicago coronata	T. hirtum
M. rigidula	T. glomeratum
M. radiata	T. fragiferum (perenn.)
M. orbicularis	T. repens (perenn.)
M. minima	T. subterraneum
M. truncatula	T. brachycalycinum
M. rugosa	T. nigrescens
M. polymorpha	T. yanninicum
M. littoralis	Trigonella mesopotamica
M. tornata	T. foenum-grecum
M. scutellata	T. monspeliosa
M. sativa (perenn.)	T. monantha
M. arabica	T. stellata
M. murex	T. asroites
Onobrychis christa-galli	Vicia narbonensis
O. viciifolia (perenn.)	V. sativa
O. kotschyana (perenn.)	V. villosa
Ononis reclinata	V. amphicarpa

Phosphate Fertilization of Mediterranean Pastures

Phosphate fertilization of annual pasture is widely used in the Mediterranean-type climates of California and Australia. In both cases however, the practice is closely linked with the seeding of the pasture with introduced legume species (annual medics and subterranean clover). This is due to the lack of native herbaceous legumes in the case of Australia (Donald 1970) and the poor productivity of native legumes in California (Jones 1974; Jones et al. 1970; Martin and Berry 1956; Nickell et al. 1962; Williams et al. 1956). An important consequence of phosphate fertilization in Australia was the sharp increase in livestock production as a result of improved pasture productivity (Russell 1960). Phosphate fertilization of marginal lands in the Mediterranean basin itself is not widely used, although, because of the presence of native legumes, there would seem to be even greater potential than existed in California and Australia. In Portugal, Spain, the south of

France, and Italy there are several examples where the use of superphosphate after scrub clearing resulted in animal output several times greater than controls (Crespo 1985).

From studies with solution cultures, Andrew (1962) indicated that the critical concentration of phosphorus in solution for optimum growth of pasture legume species lies between 0.1 and 0.2 ppm. Asher and Longeragan (1967) found a similar value (0.16 ppm) for *Trifolium subterraneum* (variety Mt. Barker). Such a concentration must therefore exist in the soil solution for the legume plants to obtain adequate phosphorus.

Studies of rangeland soils with low levels of sodium bicarbonate - soluble phosphorus were also found to have much less than 0.2 ppm in their soil solutions (Osman et al. 1977). However, the phosphate concentration which might exist in the soil solution after adding phosphate fertilizer is affected by the adsorption characteristics of the soil. Adsorption characteristics are different with different soils and can be found through analytical determination of adsorption isotherms (Beckwith 1965; Ozanne and Shaw 1967; Fox et al. 1968). In rangelands the sodium bicarbonate-soluble phosphorus in mediterranean California was rated by Murphy et al. (1973) as very low; low; intermediate; and high; when the values were: <5; 5–10; 10–20; and >20 ppm, respectively.

The rest of this paper will be devoted to a case study involving the use of phosphate fertilizer in northern Syria. The objectives of the study, at Tel Hadya (the ICARDA main research station) in north Syria, are to:

1. measure the economic impact of using phosphorus in terms of animal productivity;
2. observe the effect of stocking rate and its interaction with phosphorus on the stability of a marginal land ecosystem; and
3. determine the effect of phosphorus application on the botanical composition of grazed pastures with particular reference to its effect on legumes.

The experimental variables are 3 different fertilizer treatments, 0, 25 and 60 kg P_2O_5/ha applied in the form of triple superphosphate, every year in November; and 2 stocking rates, low (1.3 ha/sheep) high (0.6 ha/sheep). Ninety Awassi ewes were divided into groups of five, each containing 2, 3, 4, 5 and 6 year-old ewes. Each group was then permanently assigned to large (6.5 ha) or small (3 ha) plots, representing the low and high stocking rates respectively in each of the fertilizer treatments. Each year animals graze the plots from early morning to sunset and are sheltered at night. Sheep are fed with barley grain during late pregnancy and early lactation (December to February) and in July and August each year in preparation for mating. Barley and hay is also provided whenever liveweight drops below 43 kg.

Analytical results prior to the application of superphosphate fertilizer indicate that the soil is low in phosphorus (Fig. 1): over 87% of the samples analyzed had values less than 10 ppm, but the level has improved as the

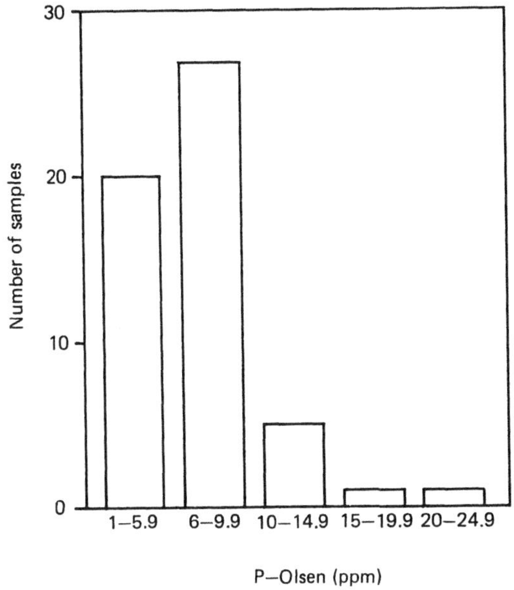

Fig. 1. Distribution of phosphorus levels (ppm) in 54 soil samples from marginal land prior to the application of superphosphate fertilizer.

result of two years of phosphate application (Fig. 2). While 83% of the samples analyzed in the control treatment (Fig. 2) had values less than 10 ppm of P, the medium and high treatments always gave values greater than 10 and 15 ppm respectively. It is clear that the use of fertilizer has corrected the phosphorus deficiency.

The effect of improving soil phosphorus on growth and productivity of native pasture legumes is profound. The data in Table 2 reflect the importance of legumes in relation to other species, the legumes increasing both as a result of the fertilizer and also with time. Grazing pressure and animal selectivity certainly play a role in the latter, as indicated by a consistently high legume percentage at low stocking rate. Annual variations in yield of legumes were observed over the years of the experiment. These are partly attributed to rainfall (amount and distribution), although there is a strong suggestion that the effect of fertilizer is accumulating each year. For example, following the first season, herbage yield doubled as a result of phosphorus application, while in the fourth year it was 2.5 to 3 times as much. Such increases could be attributed to both the build up with time of plant numbers and to the healthy plant growth (higher plant weight) under fertilizer treatments. In 1987/88 the herbage yield of legumes ranged between 2.2 and 2.8 t/ha, reflecting the potential productivity of marginal land when moisture and phosphorus are not limiting factors.

The increased plant population resulted in increased levels of organic

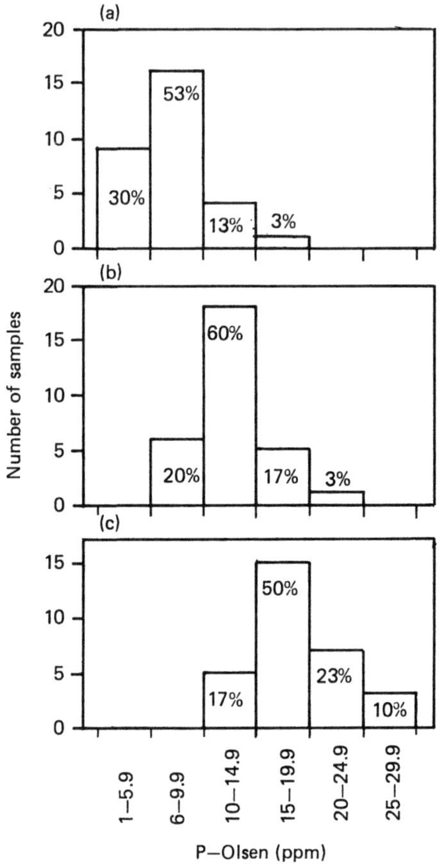

Fig. 2. Frequency distribution of available phosphorus level (P-Olsen) in marginal land soils at Tel Hadya after two years of zero (a), 12 kg/ha (b) and 28 kg/ha (c) of phosphorus application.

Table 2. Annual legumes (% composition) on marginal land in relation to level of phosphate fertilizer application and grazing pressure in different seasons.[1]

Treatment	Level	Season[2]		
		1985/86	1986/87	1987/88
	0	26.4 a	38.0 a	48.5 a
Phosphate fertilizer ($kg\ P_2O_5/ha$)	25	31.2 a	46.8 b	51.5 a
	60	29.3 a	48.7 b	52.0 a
Grazing pressure[3]	low	30.5 a	46.5 a	52.6 a
	high	27.1 a	42.4 a	48.6 a

[1] Estimated in April–May each season.
[2] Means in each column are significantly different (P = 0.01) if followed by different letters.
[3] Low = 1.3 ha/sheep; and high = .65 ha/sheep.

Table 3. Feed consumption (kg) by sheep grazing marginal land pasture as affected by phosphate fertilization and grazing pressure in 1987/88 season.

Treatment	Level	Feed consumption[1]			
		Before Lambing		*After Lambing*	
		Barley Grain	Hay[2]	Barley Grain	Hay
	0	27.7 a	8.1 a	22.5 a	21.8 a
Phosphate fertilizer (kg P_2O_5/ha)	25	15.1 ab	5.6 a	16.2 a	16.1 a
	60	9.2 b	3.8 a	20.1 a	17.5 a
Grazing pressure[3]	low	4.1 a	1.7 a	16.7 a	14.4 a
	high	30.5 b	9.9 b	22.5 a	22.6 b

[1] Each value is an average of five sheep; values in each column are significantly different (P = 0.01) if followed by different letter(s).
[2] Vetch/barley.
[3] Low = 1.3 ha/sheep and high = 0.65 ha/sheep.

matter in the soil, especially under phosphate treatments. The level of increase in organic matter was also affected by stocking rate, being highest under low stocking rate.

One of the most important factors affecting the state of health and productivity of the annual pasture is seed production, since it directly affects plant population which in turn affects pasture productivity. Seed production by annual legumes in this study was increased by 28 to 61% in the first year as the result of applying phosphate fertilizer equivalent to 25 kg P_2O_5/ha and 60 kg P_2O_5/ha, respectively. The following year increases were 56 and 93% while in the third season corresponding values were 225 and 371%. We expect this cumulative increase to continue.

Probably the most important impact of marginal land improvement is one which can be expressed in economic terms. Examples from Portugal, Spain, and the south of France and Italy have indicated the high potential for increasing animal output on marginal lands (Crespo 1985). Equally important, however, are the aspects of marginal land improvement leading to stabilization and/or control of soil erosion, even though these cannot be expressed directly in terms of economic benefits (Veronesi 1987). In the present study, although the full economic analysis is still to be done, some results can be highlighted. For example, Table 3 shows the supplementary feed required by sheep to prevent the drop in body weight of ewes below 43 kg. The sheep grazing on fertilized plots required far less barley and hay than the controls, and the differences were significant for barley before lambing. Sheep grazing at a high stocking rate required considerably higher levels of feed.

Considering the level of production of pasture at the end of April this year (Table 4), and assuming this represents only half of the available feed (total = grass + legume), one can get some idea of the potential for livestock production on marginal land. For proper pasture management, animals are allowed to graze half of the total feed available. This means that the amount shown in Table 4 can be considered available for grazing till the end of the year. Assuming 800 kg DM are required by the ewe per year (Carter 1987) the total feed needed to support one sheep for 8 months (May–December)

Table 4. Dry matter yield (kg/ha) in April 1988 of legume species on marginal land as affected by phosphate fertilization and the estimated number of sheep that can be supported on a hectare of pasture till the end of the year.

Phosphate fertilizer level (kg P_2O_5/ha)	Dry matter	Sheep numbers
0	878.5	1.65
25	2163.0	4.06
60	2831.0	5.31
LSD (P = 0.05)	543.2	

can be estimated as 533 kg. This shows that the feed produced under the two fertilizer treatments (25 kg P_2O_5 and 60 kg P_2O_5) can support 4.1 sheep/ha and 5.3 sheep/ha, respectively, till the end of the year. This is exactly 2.5 times and more than 3 times the potential stocking rate of the control treatment.

Obviously the above situation represents a good rainfall season but even in 1986/87, the additional feed was more than the animals could eat, which brought a significant reduction in the use of concentrates during summer and early winter.

References

Abd El Moneim, A.M., and Cocks, P.S. 1986. Adaptation of *Medicago rigidula* to a cereal-pasture rotation in north-west Syria. Journal of Agricultural Science, Cambridge, 107: 179–186.

Beckwith, R.S. 1965. Sorbed phosphate at standard supernatant concentration as an estimate of the phosphate needs of soils. Australian Journal of Experimental Agriculture and Animal Husbandry. 5: 52–58.

Carter, E.D. 1978. Legumes in farming systems of the Near East and North African region. Report to ICARDA; University of Adelaide, Australia.

Cocks, P.S., Thomson, E.F., Somel, K., and Abd El Moneim, A.M. 1988. Degradation and rehabilitation of agricultural land in north Syria, ICARDA-110 Ar, En, Aleppo. International Centre for Agricultural Research in the Dry Areas, Aleppo, Syria.

Crespo, D.G. 1985. Importance of grazing trials in determining the potential of rainfed Mediterranean pastures. Pages 85–92 *in* FAO-European Cooperative Network on Pasture and Fodder Crop Production, Bulletin No.4. Elvas, Portugal 16–19 April 1985.

Donald, C.M. 1970. Temperate pasture species, *in* 'Australian Grassland' (Moore, R.M. ed.), Australian National University Press, Canberra, Australia.

Fox, R.L., Plucknett, D.L. and Whitney A.S. 1968. Phosphate requirement of Hawaiian Latosols and residual effects of fertilizer phosphorus. Pages 301–310 *in* 9th International Congress of Soil Science: II. Transactions. Adelaide, Australia.

Jones, M.B. 1974. Fertilization of annual grasslands of California and Oregon *in* Forage fertilization (Mays, D.A. ed.) ASA, CSSA, SSSA. Madison, WI, USA.

Jones, M.B., Lawler, P.W., and Ruckman J.E. 1970. Differences in annual clover responses to phosphorus and sulphur. Agronomy Journal 62: 439–442.

Martin, W.E., and Berry, L.J. 1956. Fertilized range can pay dividends. Results of ten grazing tests on annual range, 1954–1955 season. Progress report. University of California Agricultural Extension Service, Berkeley, USA.

McKell, C.M., Wilson, A.W., and Williams, W.A. 1962. Effect of temperature on phosphorus utilization by native and introduced legumes. Agronomy Journal 54: 109–113.

Murphy, A.H., Jones, M.B., Clawson, J.W. and Street, J.E. 1973. Management of clovers on California annual grasslands. Californian Agricultural Experimental Station Extension Service Circular No.564, Berkeley, USA.

Osman, A., Raguse, C.A. and Summer, D.C. 1977. Growth of subterranean clover in a range soil as affected by microclimate and phosphorus availability: II. Laboratory and phytotron studies. Agronomy Journal 69: 26–29.

Ozanne, P.G., and Shaw, T.C. 1967. Phosphate sorption by soils as a measure of the phosphate requirement for pasture growth. Australian Journal of Agricultural Research 18: 601–612.

Puckridge, D.W., and French, R.J. 1983. The annual legume in cereal-ley farming systems of southern Australia: a review. Agriculture, Ecosystems, and Environment 9: 229–267.

Russell, J.S. 1960. Soil fertility changes in the long term experimental plots at Kybybolite, South Australia. I. Changes in PH, total nitrogen, organic carbon, and bulk density. Australian Journal of Agricultural Research 11: 902–926.

Veronesi, F. 1987. Perspectives for increasing productivity of pastures in central and South Italy. Pages 193–194 *in* Workshop on pasture improvement, Madrid, Spain 22–24 April 1987. Commission of the European Communities, EUR 11170 EN.

Williams, A.W., Love, R.M. and Conrad, J.P. 1956. Range improvement in California by seeding annual clovers, fertilization and grazing management. Journal of Range Management 9: 28–33.

Discussion

Papastylianou:
You refer to 'marginal lands' in your study but the work was done in areas receiving 300–500 mm rainfall. What would be the effect of phosphorus fertilizer in the real marginal lands which receive 200 mm or less?

Osman:
I gave a definition of what we mean by marginal lands as non-arable land within the cereal zone, so rainfall could be from 200 mm, where barley is grown, to 500 mm. We haven't tried using fertilizer in areas receiving less than 200 mm yet.

Gintzburger:
I think 'marginal land' is any land where you cannot cultivate, but just keep your animals. We did some similar work in Libya which supports the work at ICARDA.

Falcinelli:
Water limitation is not the only reason for classifying land as 'marginal'. Land which is too steep or stony for cultivation, with soils that are saline or skeletal, should also be termed marginal.

Buddenhagen:
In California we consider non-crop land to be 'rangeland'. 'Marginal' land is land of low productivity for whatever cause.

Osman:
May be a better term would be 'Mediterranean grassland'. 'Rangeland' in this region refers to a different category of natural resources.

Buddenhagen:
You stated that with phosphate fertilizer you could stock up to five animals/ha without supplements, yet your Table 3 shows considerable supplementary feeding even at 0.65 or 1.3 ha/animal. Please explain.

Osman:
Table 3 refers to the situation Oct-Dec 1987 (before lambing) and Feb-March 1988 (after lambing). The high feed availability after that, April-Dec 1988, suggested that no supplementary feed will be required and more sheep could be fed on the pasture in future.

Beale:
This experiment has two major factors, or management changes; (1) application of phosphate, and (2) set stocking. There is a trend in the trial for an improvement in the percentage of legumes over the years with nil phosphate. My concern is that the gains demonstrated by adding phosphate may not be made without appropriate changes in animal management. If the pasture is grazed right down the phosphorus gains are nothing. Control of grazing is not so easy.

Cocks:
In the original strategy for the experiment we knew we had to do two things, change management, and encourage legumes by adding phosphate. You can't do one without the other. We have chosen to find the technical answers first. There are clear difficulties in applying them to real-life situations.

Beale:
In Morocco there are now some fields that have excellent legume pastures. These offer almost instant cultivars or mixtures ready to be harvested and extended to other farms. For example, there is a field that has four species of medics and over 200 kg/ha of seed ready for harvest. (*M. truncatula*, *M. aculeata*, *M. orbicularis* and *M. polymorpha*). They also offer an opportunity to study the management that has resulted in these fields being rich in legumes.

Gintzburger:
What do you think about spreading phosphate every year? Is it possible with marginal land conditions? Maybe we should apply it in one dose and observe the effect over several years.

Osman:
Applying the fertilizer in one dose is the easiest and may be the cheapest method, but split application has the advantage of maximizing pasture response with time, especially when the pasture has a low legume population to begin with.

Haddad:
This is an important factor which should be considered because it affects the cost of production. If you need an average of 1 to 2 ha/sheep, as indicated in your presentation, how much will it cost a farmer with 50 or 100 sheep to apply fertilizer? Would they be able to afford to do this every year?

Osman:
The decision will depend on several factors: level of phosphorus deficiency; climatic conditions, especially annual rainfall; proportion of legumes in the sward; and the cost of fertilizer application. On the international market 1 kg of P_2O_5 costs 0.25 USD.

Kamel:
In the majority of countries in this region fertilizers are imported and highly subsidized, and government policies often do not allow farmers to have access to these fertilizers. How do you see the role of ICARDA in this respect? Should it, along with National Program scientists, try to influence government policies in such cases?

Osman:
One of the major questions we are trying to answer in this experiment is the question of economics. Animal (sheep) productivity is one of the most important factors being assessed. When the final results are in favor of changing current practices it should not be difficult to convince officials and policy makers.

Robson:
1. Is there a change in the botanical composition within the legume component with time after phosphate application?
2. Has the plant material been analyzed for P and N?
3. What is the depth distribution of the organic matter?

Osman:
1. There were three dominant legume species at the beginning of the experiment and they are still the same: *Trifolium campestre*, *T. tomentosum* and *T. stellatum*.
2. We do phosphorus tissue analysis.
3. The soil is variable in depth (1 to 50 cm) and so is the distribution of organic matter.

Capper:
To what extent are the improvements in animal efficiency which you present related to improved quantity and N content of the herbage produced, and to what extent do the results depend upon increasing the phosphorus content of the diet? Phosphorus deficiency can have profound effects on feed conversion efficiency, growth rates, fertility and milk production in ruminants.

Osman:
At this stage it is difficult to quantify the contribution of the various factors. For example, data recently collected from the experiment show improved fertility in ewes grazing on phosphate-fertilized plots. Whether this is due to more phosphorus in the diet or to more feed is difficult to say. It may be due to both factors.

Snobar:
How can you be sure that the effects you are getting are not due to phosphate from previous seasons?

Osman:
All we are trying to show is that adding phosphate, even in small amounts, every year is resulting in a build-up of phosphorus available in the soil, and that the pasture legumes are responding to this new situation. It is not our intention to differentiate between the effects of fresh application and the residual effect of phosphorus from previous years.

Djemali:
Why don't ICARDA scientists think of a strategy to improve animal production by decreasing the number of animals and increasing the productivity of those remaining?

Osman:
I think livestock numbers will decrease automatically when people have more productive breeds. The strategy should therefore rely on identification and selection of more productive animals within existing populations, and in improving animal nutrition.

Jones:
Could Dr Osman and his co-authors indicate where the work should go next:
1. In respect of existing trials; will you stop adding P and see what happens?
2. In respect of taking the work to farmer conditions; would you be prepared to take a whole area of marginal land, spread P, and see what happens?
Learning from the experience of the trial, have you any comments on how others might undertake trials of this type?

Osman:
The experiment is a long-term one (5–6 years), therefore will continue for at least two more seasons. Meanwhile it is important to gain a better understanding of the ownership and management of the communal lands, which I think is going to be the next step. On-farm activities, if decided, will be conducted in collaboration with the national programs.

The Role of Self-regenerating Pasture in Rotation with Cereals in Mediterranean Areas

A.D. ROBSON

Soil Science and Plant Nutrition, School of Agriculture, University of Western Australia, Nedlands, Western Australia 6009, Australia.

Abstract. There are considerable benefits to both cereal and animal production in including self-regenerating pasture legumes in rotations in Mediterranean areas. These benefits are (i) inputs of nitrogen for cereal and pasture production, (ii) increased levels of organic matter leading to improved soil structure, (iii) the provision of a disease break, and (iv) the integration of grazing on legume pastures and cereal residues.

There are other interactions between cereal production and pasture production which include the control of weeds in cereals by manipulation of pastures, and the effect of herbicides and management of cereal residues on the regeneration of legume pastures. The role of self-regenerating legume pastures is best evaluated using whole-farm models which take account of the interactions between enterprises.

To maximize benefits from self-regenerating pasture legumes, attention must be paid to selecting both legumes and rhizobia adapted to the climatic and edaphic environments. Particular attention must also be paid to methods of establishment and fertilization of the pastures, and to the effect of tillage, herbicides and stubble management on regeneration.

Introduction

The inclusion of self-regenerating pasture legumes in rotations in Mediterranean areas can significantly increase both cereal yields and animal production (Puckridge and French 1983). Because of the complex interactions between the cropping and animal enterprises, benefits of including leguminous pastures in rotations with cereals are best determined using whole-farm models (Morrison et al. 1986; Kingwell and Pannell 1987). However, because a central feature of these models is quantifying the interactions between enterprises, in this review I will consider initially the effect of pastures on crop production, and the effect of crops on pasture and animal production. Subsequently I will examine how self-regenerating pasture legumes can be managed in order to obtain the maximum benefits for the whole rotation.

A.E. Osman et al. (eds.), The Role of Legumes in the Farming Systems of the Mediterranean Areas, 217–236.

Interactions Between Pastures and Crops

Interactions between pastures and crops arise mainly through effects on soil fertility. Soil fertility can be defined as the ability of the soil to produce a marketable commodity in a particular environment (Hallsworth 1969). Soil fertility has many components which can be grouped into three broad categories – biological, physical, and chemical (Table 1). Effects of pastures on cereal production and effects of crops on pasture production can operate through effects on several components. Sometimes the effects of agricultural practices on the components of soil fertility are synergistic; sometimes they are antagonistic. For example, on the one hand, reduced tillage may increase cereal yields by improving soil structure (Hamblin and Tennant 1981), on the other, it may decrease cereal yields by encouraging the incidence of diseases such as *Rhizoctonia* (Jarvis and Brennan 1986), or by decreasing the availability of nitrogen (Jarvis et al. 1985) and phosphorus (Cornish 1987). In any one environment the effect of introducing reduced tillage will depend upon the relative importance of these opposing effects. To understand and to predict the effects of pastures on cereal yields and the effects of cereal on pasture production, we need to examine the effects of crops and pastures on all the components of soil fertility.

There may also be interactions between cereal production and pasture production associated with (i) the grazing of cereal stubbles by sheep, (ii) controlling weeds of cereal crops by manipulation of pastures, (iii) the effect of herbicides applied to control weeds in cereal crops on the regeneration of pastures, and (iv) effects of stubble management on the regeneration of pastures.

Table 1. Components of soil fertility.

Chemical
Nutrient supply
Acidity
Salinity
Physical
Soil strength
Surface crusting
Factors affecting water infiltration
Porosity
Biological
Pathogenic fungi
Soil animals
Mycorrhizal fungi
Rhizobia
Microorganisms involved in nutrient cycling

Effect of Pastures on Crop Production

Inclusion of self-regenerating legumes into rotations can affect cereal production by (i) increasing the availability of nitrogen for cereal crops, (ii) improving the soil physical structure, (iii) increasing soil acidity, (iv) increasing water repellence in sandy soils, (v) providing a 'disease break', and (vi) affecting the weed population. Each of these factors will be considered in turn.

Nitrogen Availability

Nitrogen input by legumes is often likely to be the most important factor in increased cereal yields in crops following pasture legumes. However, the role of legumes in improving cereal yield can not always be substituted for by application of nitrogenous fertilizers (Table 2). Increased yields in the absence of nitrogenous fertilizers were not associated with equal increases in water use efficiency (Hamblin et al. 1987).

Symbiotic nitrogen fixation by legumes is extremely important, not only as a source of nitrogen for subsequent cereal crops but also to increase the production and utilization of pastures. Many soils are deficient in nitrogen for the growth of non-legumes in pastures, even after the growth of effectively nodulated legumes for several years (Cocks 1980). In Mediterranean environments, the higher nitrogen concentrations in legume residues, including seeds, compared with non-legume residues at the end of the season (Gladstones and Loneragan 1975) are important in the utilization of these residues by animals over the summer and autumn.

Most estimates of nitrogen input under legume-based pastures are derived by analysis of soils for total nitrogen after varying periods of pastures (Ladd and Russell 1983). These estimates of inputs (40 to 160 kg N/ha/year) do not accurately assess the symbiotic nitrogen fixation because little attempt is generally made to estimate either rates of non-symbiotic nitrogen fixation, or rates of loss of nitrogen by volatilization or leaching. Also, inputs are frequently estimated from experiments in which the pastures are either not grazed or 'crash-grazed'. Finally, in most experiments the inputs have been

Table 2. Water use efficiency (WUE) and grain yield (GY) for wheat grown after wheat and ungrazed subterranean clover on a deep sandplain soil at two levels of applied fertilizer nitrogen.

Previous species	Nitrogen applied (kg N/ha)	WUE (kg/ha/mm)	GY (t/ha)
Wheat	0	7.9	1.8
	335	11.9	2.9
Subterranean clover	0	13.1	3.3
	335	12.5	2.9

Source: Hamblin et al. (1987).

measured in the establishment phase of the pasture when there is a high proportion of effectively nodulated legumes in the sward.

It is likely that the amount of nitrogen fixed by pasture legumes in mixed swards will be approximately proportional to the amount of legume growth. The proportion of the total nitrogen in legumes of established legume-grass swards which is derived from soil mineral nitrogen is always less than 20% and generally less than 10% (Vallis et al. 1977; Bergersen and Turner 1983). In general, pasture legumes compete poorly with grasses for soil mineral nitrogen, although relatively few legumes have been studied. Thus in situations where there are limitations to nodulation and nodule function, the legume content of mixed swards is likely to be low.

Nodulation and nitrogen fixation of pasture legumes may be limited by several soil and climatic factors. First, introduced rhizobia may fail to colonize the soil during the growing season and fail to persist over hot, dry summers in sufficient numbers to nodulate regenerating legumes (Chatel and Parker 1973 a,b). Second, introduced effective rhizobia may fail to compete with indigenous ineffective rhizobia in forming nodules. Third, there may be deficiencies of mineral nutrients (molybdenum, copper, cobalt and calcium) directly involved in symbiotic nitrogen fixation (Robson 1983). Fourth, nodulation and nitrogen fixation may be more sensitive than the growth of the host legume to environmental stresses such as acidity and salinity (Robson and Loneragan 1970; Robson 1988), water stress (Sprent 1976), and low temperatures (Robson and Abbott 1987).

Management factors which increase the proportion of legumes within pastures or pasture production while maintaining the same proportion of legumes are also likely to increase nitrogen inputs by symbiotic nitrogen fixation. For example, the marked association between total soil nitrogen and extractable soil phosphorus observed in a survey of forty-four fields with varying histories of treatment with subterranean clover and superphosphate (Donald and Williams 1954), probably directly reflects effects of phosphorus in increasing legume production. Initial application of phosphorus may increase the proportion of clover in the pasture (Willoughby 1954). However, continued application of phosphorus generally leads to non-legume dominance (Rossiter 1966), reflecting an increased nitrogen supply because of the initial increase in legume production. In narrow rotations of legume pastures with cereal crops the move to non-legume dominance may not occur because of the utilization of the fixed nitrogen by the cereal crop.

The long-term effect of including legume pastures in rotations with cereals on in soil nitrogen will depend upon (i) the relative frequency of pasture years and crop years (Greenland 1971), (ii) the amount of legume in the pasture, (iii) the yield of the cereal which affects nitrogen removal, (iv) the initial nitrogen status, and (v) cultivation practices. In general, changes in soil nitrogen with either pasture or crop can be described using the equation $dN/dt = -KN + a$ where ‘K’ is a decomposition constant, N is soil nitrogen content at time ‘t’, and ‘a’ is an accretion constant. For legume pastures a mean value for 'a' could be 80 kg/ha/year and 'K' has been estimated to be 0.01. Under cereal crops 'a' has been estimated to be 20 kg/ha/year and 'K' to be 0.04 (Greenland 1971). Using these values a 1:1 pasture-wheat rotation would maintain soil nitrogen at 0.10%.

Fallowing after legume pastures which increases the mineralization of organic nitrogen (Tuohey et al. 1972; French 1978) may also lead to greater losses of nitrogen by either leaching or denitrification. Russell (1981) has estimated that decomposition coefficients of soil nitrogen under fallow are almost twice those under wheat. Where nitrogen is not lost by leaching, fallowing after medic pastures to conserve water may mineralize excessive nitrogen leading to haying-off of cereal crops (Tuohey et al. 1972).

It is likely that soil nitrogen will increase more rapidly with pasture legumes than with grain legumes because of the greater product removal of nitrogen with grain legumes. As with pasture legumes, nitrogen inputs by grain legumes appear to be closely related to the amount of legume growth and the availability of mineral nitrogen (Evans and Herridge 1986). Estimates of nitrogen fixation by grain legumes are perhaps greater than those for pasture legumes. However, because between 40 and 60% of the nitrogen in shoots is removed in seeds and the proportion of nitrogen in shoots derived from nitrogen fixation is between 40 and 90%, there will only be small gains in soil nitrogen under grain legumes in many situations.

Physical Structure of Soil

The inclusion of legume pastures in rotation can influence components of soil other than the supply of nitrogen, in particular, the soil's physical structure. The effect of pasture legumes in increasing the level of water-stable aggregates closely parallels changes in total soil nitrogen (White et al. 1978; Rowland et al. 1984). Legume pastures (i) increase porosity, (ii) increase aggregate and micro-aggregate stability, (iii) decrease bulk cohesive strength, (iv) increase water retention at any given suction, and (v) increase the amount of available water (Greenland 1971). The magnitude of the changes differs between soils and with pasture composition and vigour. The changes in the physical condition of the soil are due largely to increased biological activity including increased root growth, increased microbial activity, and increased activity of soil animals. Proliferation of roots and their associated hyphae of vesicular-arbuscular (VA) mycorrhizal fungi is important in promoting the formation of water-stable aggregates (Tisdall and Oades 1980). Root growth under pastures is generally greater than under crops. Microbial degradation of plant residues can lead to the formation and stabilization of aggregates by the production of polysaccharides that bind clay particles together (Lynch and Bragg 1985). This is probably the most important mechanism by which pastures affect aggregate stability.

The role of soil animals including earthworms in improving and maintaining soil fertility requires further research. However, the biomass of earthworms in soil in Mediterranean areas is affected by both rotation and tillage. There were almost no earthworms in soils cropped in a fallow-wheat rotation (Barley 1959). After two years of legume-based pasture there was almost as much earthworm biomass as in soils under permanent pasture. However, fallowing after legume pasture markedly decreased earthworm biomass. In another study earthworm biomass in soils sown to wheat after a subterranean clover pasture was almost twice that in soils sown to wheat following a lupin crop (Rovira et al. 1987). In this study, earthworm biomass in soils sown by direct drilling without cultivation was almost twice that in soils sown to wheat after three cultivations to 7 cm with a scarifier. The effects of both tillage and rotation on earthworm biomass probably reflects changes in the availability of substrates for earthworm growth associated with changes in the organic matter content of soil.

The effect of pastures in improving soil structure on subsequent pasture and cereal production will depend upon the initial structure of the soil. On well-structured, self-mulching clays effects of medic pastures in increasing cereal yields could be eliminated completely by nitrogen application to crops grown in a fallow-wheat rotation (Tuohey and Robson 1980). On hard-setting soils, high rates of nitrogen application (100 kg N/ha) did not restore yields of wheat on 'old crop' land (fallow-cereal rotation) to the levels obtained on land previously under pasture (Greenland 1971). On these red-brown earths, previous studies had demonstrated correlations between grain yield, seedling emergence and apparent density of the surface soil (Millington 1959). On some soils gypsum application may be required to improve soil structure to permit the establishment of vigorous legume pastures. Subsequent cropping of these soils with excessive cultivation may decrease the level of organic matter in the soil and eliminate the benefits of both gypsum application and legume pastures.

Soil Acidity

The inclusion of legume-based pastures in cereal rotations may lead to (i) soil acidification associated with increased levels of organic matter, (ii) greater removal of cations than anions, and (iii) the leaching of nitrate produced from the mineralization of fixed nitrogen (Helyar and Porter 1988). However, the growth of cereals with nitrogenous fertilizers may also lead to soil acidification the extent of which will depend on the form of nitrogen applied. It is greater with ammonium-based fertilizers than with nitrate fertilizers.

The effect of these processes on soil pH and plant growth will depend upon the initial pH of the soil and the buffering capacity of the soil for protons. In Mediterranean areas with soils containing free calcium carbonate, acidifying processes will have little effect on either soil pH or plant growth. However, in some Mediterranean areas with initially moderately acid soils, acidifying processes have decreased soil pH to levels where aluminium and manganese toxicities now either limit the species grown or decrease production (Helyar 1987).

The growth of legumes on soils low in clay content (<5% clay) may lead to the development of water repellency. Soils under pasture legumes developed more repellency than those under lupins or cereals (McGhie 1980; Summers 1987). For a wide range of sites carrying pasture, predominantly capeweed (*Cryptostemma calendulacea*) and subterranean clover, water repellency increased with increasing organic matter content (Summers 1987). However, in 1:1 rotations of wheat and subterranean clover there was no marked development of water repellency.

Disease Break

The effect of legume pastures on the incidence and severity of disease in subsequent cereal crops appears to have been investigated for relatively few diseases. However, for two major root diseases of cereals in Mediterranean areas, (cereal cyst nematode, and take-all associated with the pathogen *Gaeumannomyces graminis* var. *tritici*), inclusion of legume pastures in rotations may decrease the extent of disease. For cereal cyst nematode, two years of medic pasture decreased the number of eggs from 1.4/g soil to only 0.2/g soil (Meagher and Rooney 1966). The build-up of cereal cyst nematode is associated closely with the frequency of cropping the land with susceptible cereals (King et al. 1982).

For take-all the effect of legume pastures on incidence and severity is more complex. Grasses in legume-based pastures can act as a host for the pathogen. In one study with varying stocking rates and cultivars of subterranean clover, the incidence of take-all in cereal crops was closely positively correlated with the dry matter production of grasses in the preceding pasture phase (MacNish and Nicholas 1987). Herbicide application during the mid-winter of the pasture phase, which decreased the proportion of grass in the pasture from approximately 25 to less than 10% in mid-spring and decreased the total production of pasture by 10%, reduced the incidence of take-all in the following wheat crop from 61 to 38% (W. Macleod, unpublished data). Chemical removal of grasses from pasture leys may have marked effects on pasture and sheep production, and these will be discussed below.

The incidence of *Rhizoctonia* barepatch was similar in the following rotations: one year of medic and one year of wheat, two years of medic and two years of wheat, and three years of medic and one year of wheat (MacNish 1986). The pathogen involved in this disease appears to have a wide host range.

The effect of including legume pastures in rotation with cereals on the incidence of foliar diseases has received less attention. However, because many of the pathogens involved are either biotrophic or have narrow host ranges, it is likely that legume pastures will decrease the incidence and severity of foliar diseases.

Weed Control

Crop-pasture rotations provide greater flexibility in weed management than do long-term monocultures. Pasture management can decrease the incidence of weeds in cereals (Reeves and Smith 1975; Weels 1969). Cultivation and application of herbicides during the cereal phase can also decrease the incidence of weeds in the following pasture phase. During the pasture phase there may be a build-up of species that are weeds in subsequent crops. It is likely that the number and species of weeds in a crop of a crop-pasture rotation will depend upon the length of the preceding pasture phase. There is a need to develop integrated systems of weed control in crop-pasture rotations involving grazing and cultivation, as well as herbicide application during both the crop and pasture phases.

Effect of Crops on Pasture Production

Cropping may affect pasture production in at least two major ways: (i) cultivation and herbicide application during the cropping phase may decrease the weight of germinable seed. In some situations, herbicides which are toxic to pasture legumes may persist into the pasture phase; (ii) This is associated with stubble management. Grazing of cereal stubbles rather than pastures during summer and early autumn may permit better establishment of legume-based pastures by saving seed reserves from being eaten by animals, and by deferring grazing of germinating pastures. In some instances, large amounts of cereal residues may reduce germination of pasture legumes.

Seed Regeneration

The quantity of germinable seed, and thus plant establishment, is the major factor affecting the regeneration of legume pastures after cropping. There is also a close association between seedling density and production during winter, which is a key time in many Mediterranean areas (see Ewing and Pannell 1987).

The quantity of germinable seed depends on the level of seed production, and on the proportion of that seed that is (i) sufficiently near to the surface to emerge (within the top 2–3 cm for subterranean clover; Taylor 1985) and (ii) able to germinate, i.e. not prevented by coats impermeable to water or by physiological dormancy. Most pasture legumes grown in Mediterranean areas produce seed with coats impermeable to water. The seed coats soften with fluctuating temperatures over the summer. The extent of 'hard-seededness' varies with genotype and the conditions under which the seed is produced and over-summered. The softening process in subterranean clover appears to have two phases (Taylor 1981): an initial pre-conditioning phase in which constant temperatures may be just as effective as alternating temperatures, and a subsequent phase which is most effectively achieved with fluctuating temperatures.

Tillage Practices

In crop-pasture rotations, seeds formed by pastures are likely to be subjected to at least one summer on the surface. When the soil is cultivated for cropping, seeds which have softened will germinate and are destroyed. For subterranean clover more than half the total seed set will be destroyed, whereas for annual medics approximately one-quarter will be lost. The hard seeds remaining will be buried to various depths depending upon the tillage practices used. In one experiment, the proportion of subterranean clover seed in the surface 2 cm. was (i) 100% with no cultivation at all; (ii) 55% with a single pass by a scarifier; (iii) 35% with one disc ploughing followed by two passes by a scarifier. With ploughing and scarifying more than 20% of the total seed was between 6 and 10 cm from the surface (Taylor 1985). Burial of seeds of both subterranean clover (Taylor 1985; Taylor and Ewing 1988) and two species (*Medicago truncatula*, *M. polymorpha*) of annual medics (Taylor and Ewing 1988) markedly decreased the rate of softening of seed, probably because of the smaller fluctuations in soil temperatures with depth. Subsequent cultivation may uplift these buried hard seeds which soften rapidly and germinate. The smaller effect of burial in decreasing the softening rates of the annual medics than for subterranean clover (Taylor and Ewing 1988) suggests that more medic seeds will be lost by softening and germination at depth. There were also differences among genotypes of subterranean clover in the effect of burial on the softening of hard seeds (Taylor 1984). Seeds buried deep in the soil also suffered some microbial decomposition. While on average this represented only 3% of the total seeds, there appeared to be microenvironments within the soil where losses due to this mechanism were much greater.

If seedlings of pasture legumes are not destroyed by cultivation before sowing the crop, the seedlings may be killed by herbicides applied in the crop phase. Some herbicides in some situations (for example, chlorsulfuron on alkaline soils) may persist beyond the cropping year and prevent the regeneration of sensitive species such as medics (Reeves 1987).

Tillage may also influence pasture regeneration by affecting the distribution of rhizobia within the soil. For example, cultivation which mixes less acidic surface soil with the rest of the soil may reduce rhizobial numbers (Coventry et al. 1985). Crop establishment using either a cultivated seed bed or direct drilling decreased the pH of the surface soil and rhizobial numbers.

Cereal Residues

Medic plant density decreased exponentially with increasing mass of residual straw, so that with 4 t/ha of residual straw plant density was only one-quarter of that with 0.5 t/ha of residual straw (Quigley and Carter 1985). Cereal straw could decrease the establishment of pasture legumes by reducing soil temperature fluctuations and hence the softening of hard seeds. Cereal straw could also decrease establishment by being a physical impediment to the emerging legume seedlings or because the products of decomposition of cereal straw may be toxic to pasture legumes (Carter 1987). The levels of cereal residues associated with a grain yield of 2 t/ha assuming that 75% of the stubble is consumed by grazing animals, is approximately 1 t/ha. At this level it is likely that effects of stubble on hard seed breakdown will be the most important effect.

Management of Pasture Legumes in Crop-pasture Rotations

Selection of Appropriate Genotypes of Plant and Rhizobia

Both the pasture legumes and their associated rhizobia must be well adapted to edaphic and climatic environments and be able to colonize and persist within the constraints imposed by the rotation and grazing management.

Both the naturalized distribution of species and the success of sown legumes indicate that there is a marked association between the growth of a particular legume species and soil properties. The clearest example of this with Mediterannean pasture legumes is the difference between *Trifolium* and *Medicago* species in soil preferences (Robson 1969). *Medicago* species are restricted in their distribution to only moderately acid (>pH 6; 1/5 soil/water) and alkaline soils primarily because their rhizobia are unable to colonize and persist in more acid soils. By contrast, subterranean clover is restricted to moderately acid soils (pH 5.0–6.5; 1/5 soil/water), perhaps because of an inability to obtain mineral nutrients (phosphorus, copper and zinc) on more alkaline soils (Robson 1969). Within both *Medicago* and *Trifolium* there is also marked variation in ability to succeed in relation to soil pH. For example, the sub-species *brachycalycinum* of *Trifolium subterraneum* is better able to grow in neutral and moderately alkaline soils than are the *yanninicum* and *subterraneum* sub-species (Higgs 1958). Within the annual medic species, *M. murex* and *M. polymorpha* are able to persist better on moderately acid soils than *M. truncatula* and *M. littoralis* because of a greater capacity to form nodules under these conditions (Howieson and Ewing 1986).

Apart from soil pH and its associated soil properties, there are differences among Mediterranean pasture legumes in their response to edaphic environments. For example, *M. littoralis* and *M. tornata* appear to be better adapted to sandy soils than *M. truncatula*. Another example is the marked ability of the *yanninicum* sub-species of subterranean clover to withstand waterlogging compared to the *Brachycalycinum* and *subterraneum* sub-species (Francis and Devitt 1969).

There are also important differences among rhizobia in their responses to the edaphic environment. Chatel and Parker (1973a, b) demonstrated that strains of *Rhizobium trifolii* differed greatly in their ability to colonize and persist in sandy soils. Similarly, there is marked variation among strains of *R. meliloti* in their ability to colonize and persist in acid soils (Robson and Loneragan 1970; Howieson and Ewing 1986; Howieson et al. 1988).

Response to Phosphate

Apart from these well-demonstrated associations between soil properties and performance of annual legumes there may also be scope for selecting genotypes well-adapted to a low supply of nutrients, particularly phosphate. Within subterranean clover there appears to be considerable intra-specific variation in ability to grow well at low phosphorus supply, associated with differences in phosphate uptake rather than with differences in phosphorus utilization within the plant (Robson and Collins, unpublished data). Cultivars of subterranean clover which had been commercially successful in Australia were all among the group that showed least response to applied phosphate. In a different approach Cocks and his colleagues have examined the effect of varying phosphate supply on the shifts in composition within mixed populations of subterranean clover within long-term experiments (personal communication).

Response to Climate

Annual pasture legumes also differ in their response to the climatic environment. In dry Mediterranean environments a major factor in persistence and productivity is the capacity to produce seeds, which is largely related to maturity (Rossiter 1966). However, seed production may vary among genotypes for reasons other than variation in maturity. Annual species of *Medicago* varied in their capacity to produce flowers and in the number of pods set per flower produced (Cocks 1987). Flower survival could be predicted by counting the number of flowers at each node and weighing the pods. Annual legumes also differ in their sensitivity to water stress during flowering. A water stress imposed 14 days after the start of flowering for 21 days decreased seed yields in subterranean clover but not in annual medics (Wolfe 1985).

There are several examples of attempting to introduce pasture legumes into climatic environments in which they were poorly adapted. For example in Spain, naturalized subterranean clovers proved superior because they were earlier and had higher hard seed contents than the Australian cultivars which were introduced because of their success in regions with similar annual rainfall (Francis et al. 1977). Similarly, *M. rigidula* which is widely distributed within the Mediterranean regions with cold winters has much greater tolerance of frost than Australian cultivars of *M. truncatula*, *M. polymorpha* and *M. scutellata* (Abd El Moneim and Cocks 1986; Cocks and Ehrman 1987). Clearly, in selecting pasture legumes for inclusion in cropping systems in Mediterranean areas, particular attention must be paid to maturity, extent of hard seed production, and frost tolerance.

Rotations and Grazing Management

As well as edaphic and climatic limitations there may be limitations associated with the choice of rotation and grazing management. Thus it is likely that all commercial cultivars of subterranean clover have an insufficient proportion of hard seeds to persist in 1:1 clover: wheat rotations in low rainfall (<300 mm) areas. By contrast, annual medics, because of their higher proportion of hard seeds, may be able to persist in such narrow rotations. Subterranean clover may only have a place in ley farming systems which have an occasional cereal crop after several years of pasture.

There may also be differences among annual legumes in their ability to withstand the depletion of the seed bank by summer grazing. Survival of seed after pen feeding of sheep with medic pods in a standard lucerne chaff diet varied from 20% in a cultivar of *M. polymorpha* to 1.6% in a genotype of *M. turbinata* (Carter 1980). It may be worthwhile to select for species or genotypes with small seeds. Survival after ingestion declined exponentially with increasing seed size (Thomson et al. 1987). Small seeds would, however, need to be retained closer to the surface for successful plant establishment (see earlier discussion on effect of cultivation). Additionally, the value of annual legumes in crop-pasture rotations is at least partly related to their ability to provide high-nitrogen feed over the summer. Clearly this requirement must be closely balanced with the need for enough seed to ensure sufficient density for high winter production during the pasture phase. The use of hard-seeded legumes coupled with prior burial of the pods may permit both regeneration and intensive grazing in particular seasons when there is a late break to the growing season.

In several instances, attempts to introduce pasture legumes successful in one Mediterranean region into another region have failed. The selection of legumes suitable for a particular region involves three stages. First, an assessment of the genetic resources within the region must be made. Second, there must be a study of the distribution of indigenous species in relation to quantitative descriptions of the edaphic and climatic environment. Finally, the ecotypes identified as meeting the edaphic and climatic requirements need to be evaluated for their ability to succeed within rotations with cereals.

Establishment and Fertilization of Legume Pastures

The establishment and fertilization of legume-based pastures in Mediterranean areas have been extensively reviewed (see Carter 1987; Puckridge and French 1983; Gartrell and Bolland 1987). In this review I will consider only three aspects: (i) comparison of sowing pasture legumes with other crops (undersowing) versus sowing them on their own; (ii) selection of appropriate rhizobia; and (iii) the phosphate fertilization of pastures, particularly the role of VA mycorrhizas.

Undersowing

Pasture legumes can be sown on their own or they can be sown at low seeding rates as a mixture with a cereal crop (undersowing). While undersowing with the cereal crop requires less investment than sowing pasture legumes on their own, there are several disadvantages. Competition between the cereal crop and the pasture legume may severely decrease both grain yield and seed yield of legumes (Brownlee and Scott 1974; Ewing 1986). Additionally, undersowing cereals with legumes may limit weed control within the crop. Thirdly, as Ewing (1986) has pointed out, low seeding rates used in undersowing may lead to poor distribution of rhizobia throughout the soil with subsequent inadequate nodulation in the pasture phase. Finally, pasture sown without a cover crop can become productive in its first or second year whereas that sown under a cover crop takes longer to reach satisfactory densities (Brownlee and Scott 1974). In narrow rotations of pastures with cereals a large seed yield in the first pasture year would appear to be essential to ensure dense swards in subsequent pasture phases.

Selection of Rhizobia

A second important area in the introduction of pasture legumes into cereal rotations is the choice of rhizobial strains which form effective associations and which are saprophytically competent and, if necessary, competitive with indigenous rhizobia. There is considerable specificity among annual medic species in their ability to form effective associations with rhizobial strains (Vincent 1974; Materon and Brockwell 1987). The marked legume-rhizobial strain interaction may lead to considerable difficulties in the evaluation of legumes. Perhaps initially legumes should be evaluated for agronomic characteristics with adequate mineral nitrogen. Subsequently, their capacity to nodulate and fix nitrogen could be evaluated with a wide range of rhizobial strains including some collected at the same time as the plant. In areas where there are no naturalized legumes, particular care should be taken to introduce suitable rhizobia with the sown legume because large numbers of ineffective naturalized rhizobia may subsequently preclude the introduction of more effective strains.

Phosphate Fertilization

Adequate phosphate supply is important to the establishment of legumes in most Mediterranean areas. Pasture legumes appear to have greater external requirements for phosphorus than cereals or grasses (Abbott and Robson 1984). Hence the residual value of phosphate applied to preceding cereal crops may be insufficient for maximum growth and thus maximum nitrogen inputs by legume pastures. There are several approaches to increasing plant growth at sub-optimal phosphate supply: for example, nutrient efficient genotypes can be selected, or the symbiosis between VA mycorrhizal fungi and pasture legumes can be managed (see reviews by Abbott and Robson 1987; Robson and Abbott 1987). Many pasture legumes have relatively coarse root systems with relatively few long root hairs. Because mycorrhizas increase phosphate uptake primarily by shortening the distance phosphate must diffuse in soils, mycorrhizal legumes may require only half as much phosphate for maximum growth as non-mycorrhizal legumes (Abbott and Robson 1984).

VA mycorrhizal fungi occur in all soils, but is the level of infection adequate for the maximum benefit to be obtained from the symbiosis? In our current approach (Abbott and Robson 1987) we are attempting to predict (i) the response of plant growth in a particular soil to phosphorus supply and mycorrhizal infection; (ii) the relationship between level of infection and benefit from mycorrhizal infection; (iii) the development of mycorrhizal infection in the field using a bioassay in the glasshouse with soil collected in the summer preceding the pasture; and (iv) the effect of agricultural practices on the level of mycorrhizal infection. From this approach the benefits from either inoculation or management to increase infection can be assessed. Considerable further research is required.

Herbicide and Grazing Management

There are some opportunities to increase the proportion of legumes during the pasture phases of the rotation using specific herbicides (Venn 1984; Thorn and Perry 1987). The use of the herbicide propyzamide virtually eliminated annual grasses from a pasture and increased the growth of subterranean clover and capeweed (Thorn and Perry 1987). In the year of spraying, total pasture production was decreased, but in the following season there was no marked effect. Sheep grazing grass-free pastures had lower live-weights during winter in both years, but made compensatory gains during late spring and summer associated with the higher nitrogen concentrations in the herbicide-treated pastures. There was no effect of grass removal on wool production. In other studies removal of grasses has decreased wool growth rates (Dunlop and Thorn 1984). Removal of grasses in the legume phase has advantages in the cereal phase related both to the incidence of take-all and to the control of grass weeds within the crop.

In grazing legume pastures over summer, care must be taken to ensure that sufficient seed is left for regeneration of the pasture. Only about 2% of the seed of most commercial medic cultivars and less than 1% of the seed of subterranean clover survives passage through the sheep (Carter 1980; Carter and Lake 1985). It has been estimated that approximately 400 kg seed/ha is required for effective regeneration of pastures. Ewes grazing medic pastures gained weight as long as pod availability exceeded 10 kg/ha (Cocks 1988). There is obviously a need to monitor grazed medic pastures over summer to assess the availability of seed.

Conclusion

The introduction of pasture legumes into rotations with cereals can significantly increase both cereal and animal production in Mediterranean areas. Because of the great variation in edaphic and climatic environments and in management systems within Mediterranean areas, it is likely that the most appropriate legumes and the most appropriate procedures for maximizing benefits will vary greatly within Mediterranean areas. The genetic variation that exists within the annual species of *Medicago* is substantial and has not been adequately evaluated or exploited. There has been relatively little research to develop procedures for establishing and maintaining self-regenerating legumes in cereal rotations in Mediterranean areas other than southern Australia. Particular attention needs to be directed towards optimizing the symbiosis between legumes and rhizobia, and between legumes and VA mycorrhizal fungi.

References

Abbott, L.K. and Robson, A.D. 1984. The effect of VA mycorrhizae on plant growth. Pages 113–130 *in* VA Mycorrhizae (Powell, C. Ll. and Bagyaraj, D.J., eds). CRC Press, Boca Raton, Florida, USA.

Abbott, L.K. and Robson, A.D. 1987. Managing VA mycorrhizal fungi in pastures. Pages 10–12 *in* 7th North American Conference on Mycorrhizae (Sylvia, D.M., Hung, L.L. and Graham, J. H. eds). University of Florida, Gainesville, USA.

Abd El Moneim, A.M. and Cocks, P.S. 1986. Adaption of *Medicago rigidula* to a cereal-pasture rotation in north-west Syria. Journal of Agricultural Science (Cambridge) 107: 179–186.

Barley, K.P. 1959. Earthworms and soil fertility. 1. Earthworm populations found in agricultural land near Adelaide. Australian Journal of Agricultural Research 10: 171–178.

Bergersen, F.J. and Turner, G.L. 1983. An evaluation of ^{15}N methods for estimating nitrogen fixation in a subterranean clover-perennial ryegrass sward. Australian Journal of Agricultural Research 34: 391–401.

Brownlee, H. and Scott, B.J. 1974. Effects of pasture and cereal sowing rates on production of undersown barrel medic and wheat cover crop in western New South Wales. Australian Journal of Experimental Agriculture and Animal Husbandry 14: 224–230.

Carter, E.D. 1980. The survival of medic seeds following the ingestion of intact pods by sheep. Page 178 *in* Proceedings of the Australian Agronomy Conference, Lawes, Queensland. Australian Society of Agronomy.

Carter, E.D. and Lake, A.W.H. 1985. Seed, seedling and species dynamics of grazed annual pastures in Southern Australia. Pages 654–656 *in* Proceedings of the XV International Grassland Congress, Kyoto, Japan. Japanese Society of Grassland Science, National Grasslands Research Institute, Japan.

Chatel, D.L. and Parker, C.A. 1973a. Survival of field grown rhizobia over the dry summer period in Western Australia. Soil Biology and Biochemistry 5: 415–423.

Chatel, D.L. and Parker, C.A. 1973b. The colonization of host root and soil by rhizobia. 1. Species and strain differences in the field. Soil Biology and Biochemistry 5: 425–432.

Cocks, P.S. 1980. Limitations imposed by nitrogen deficiency on the productivity of subterranean clover-based annual pasture in Southern Australia. Australian Journal of Agricultural Research 31: 95–107.

Cocks, P.S. 1987. Flower production and seed survival on spaced plants of annual medics. Pages 105–123 *in* Annual Report of Pasture, Forage and Livestock Program, ICARDA, Aleppo, Syria.

Cocks, P.S. 1988. Seed production and seed survival under grazing of annual medics (*Medicago* spp.) in north Syria. Journal of Agricultural Science, Cambridge 110: 455–463.

Cocks, P.S., and Ehrman, T.A.M. 1987. The effect of geographic origin on host tolerance of pasture legumes in Syria. Journal of Applied Ecology 24: 678–683.

Cornish, P.S. 1987. Effects of direct drilling on the phosphorus uptake and fertilizer requirements of wheat. Australian Journal of Agricultural Research 38: 775–790.

Coventry, D.R., Hirth, J.R., Reeves, T.G. and Jones, H.R. 1985. Changes in soil populations of *Rhizobium trifolii* following crop establishment in a clover ley rotation in south-east Australia. *In* Proceedings of the XV International Grasslands Congress, Kyoto, Japan. Japanese Society of Grassland Science, National Grasslands Research Institute, Japan.

Donald, C.M. and Williams, C.H. 1954. Fertility and productivity of a podsolic soil as influenced by subterranean clover (*Trifolium subterraneum* L.) and superphosphate. Australian Journal of Agricultural Research 5: 664–687.

Dunlop, A.C. and Thorn, C.W. 1984. Animal production from grass-free annual pastures. Page 673 *in* Proceedings of the Australian Society of Animal Production 15.

Evans, J. and Herridge, D.F. 1986. Nitrogen inputs and utilization in crop legumes. Pages 14–43 *in* Nitrogen Cycling in Temperate Agricultural Systems (Bacon, P.E., Evans, J., Storrier, R.R. and Taylor, A.C., eds). Australian Society of Soil Science, Wagga Wagga, New South Wales.

Ewing, M.A. 1986. Serena and Circle Valley medic establishment. Journal of Agriculture, Western Australia 27: 107–112.

Ewing, M.A. and Pannell, D.J. 1987. Development of regional pasture research priorities using mathematical programming. Pages 583–585 *in* Temperate Pastures, Their Production, Use and Management (Wheeler, J.L., Pearson, C.J. and Robards, G.E., eds). CSIRO, Melbourne, Australia.

Francis, C.M. and Devitt, A.C. 1969. The effect of waterlogging on the growth and isoflavone concentration of *Trifolium subterraneum* L. Australian Journal of Agricultural Research 20: 819–824.

Francis, C.M., Ramos, M.A. and Quinlivan, B.J. 1977. Predicting the maturity requirements of subterranean clover cultivars for farm grazing in Extramadura, Spain. CSIRO Plant Introduction Review 12: 14–28.

French, R.J. 1978. The effect of fallowing on the yield of wheat. I. The effect on soil water storage and nitrate supply. Australian Journal of Agricultural Research 29: 653–660.

Gartrell, J.W. and Bolland, M.D.A. 1987. Phosphorus nutrition of pastures. Pages 127–136 *in* Temperate Pastures – Their Production, Use and Management (Wheeler, J.L., Pearson, C.J. and Robards, G.E., eds). CSIRO, Melbourne, Australia.

Gladstones, J.S. and Loneragan, J.F. 1975. Nitrogen in temperate crop and pasture plants. Australian Journal of Agricultural Research 26: 103–112.

Greenland, D.J. 1971. Changes in the nitrogen status and physical condition of soils under pastures with special reference to the maintenance of the fertility of Australian soils used for growing wheat. Soils and Fertilizers 34: 237–251.

Hallsworth, E.G. 1969. The measurement of soil fertility: the National Soil Fertility project. Journal of the Australian Institute of Agricultural Science 35: 78–89.

Hamblin, A.P. and Tennant, D. 1981. The influence of tillage on soil water behaviour. Soil Science 132: 233–239.

Hamblin, A.P., Tennant, D. and Perry, M.W. 1987. Management of soil water for wheat production in Western Australia. Soil Use and Management 3: 63–69.

Helyar, K.R. 1987. Nutrition of plants on acid soils. Pages 159–171 *in* Temperate Pastures – Their Production, Use and Management (Wheeler, J.L., Pearson, C.J. and Robards, G.E., eds). CSIRO, Melbourne, Australia.

Helyar, K.R. and Porter, W.M. 1988. Soil acidification, its measurement and the processes involved. *in* Soil acidity and Plant Growth (Robson, A.D., ed.). Academic Press, Sydney. In press.

Higgs, E.D. 1958. Choosing subterranean clover strains for your farm. Journal of the Department of Agriculture, South Australia 61: 267–272.

Howieson, J.G. and Ewing, M.A. 1986. Acid tolerance in the *Rhizobium meliloti – Medicago* symbiosis. Australian Journal of Agricultural Research 37: 55–64.

Howieson, J.G., Ewing, M.A. and d'Antouno, M.F. 1988. Selection for acid tolerance in *Rhizobium meliloti*. Plant and Soil 105: 179–188.

Jarvis, R.J. and Brennan, R.F. 1986. Timing and intensity of surface cultivation and depth of cultivation affects *Rhizoctonis* patch and wheat yield. Australian Journal of Experimental Agriculture 26: 703–708.

Jarvis, R.J., Hamblin, A.P. and Delroy, N.D. 1985. Continuous cereal cropping with alternative tillage systems in Western Australia. Western Australian Department of Agriculture Technical Bulletin 71, Perth, Australia.

King, P., Rovira, A.D., Brisbane, P.G., Simon, A. and Brown, R.H. 1982. Population estimates of cereal cyst nematode and response of wheat to granular nematicides. Australian Journal of Experimental Agriculture and Animal Husbandry 22: 209–220.

Kingwell, R.S. and Pannell, D.J. 1987. (eds). Midas, a bioeconomic model of a dryland farm system. Pudoc, Wageningen, 207 pp.

Ladd, J.N. and Russell, J.S. 1983. Soil nitrogen. Pages 589–607 *in* Soils: An Australian Viewpoint. CSIRO Division of Soils, Melbourne Academic Press, London.

Lynch, J.M. and Bragg, E. 1985. Microorganisms and soil agregate stability. Advances in Soil Science 2: 133–171.

Materon, L.A. and Brockwell, J. 1987. Symbiotic response of medic species to indigenous rhizobia. Pages 198–200 *in* Annual Report for 1987, Pasture, Forage and Livestock Program. ICARDA, Aleppo, Syria.

MacNish, G.C. 1986. *Rhizoctonia* patch of cereals. Journal of Agriculture, Western Australia 27: 91–95.

MacNish, G.C. and Nicholas, D.A. 1987. Some effects of field history on the relationship between grass production in subterranean clover pasture, grain yield and take-all (*Gaeumannomyces graminis* var. *tritici*) in a subsequent crop of wheat at Bannister, Western Australia. Australian Journal of Agricultural Research 38: 1011–1018.

McGhie, D.A. 1980. The origin of water repellence in some Western Australian Soils. Ph.D. Thesis, The University of Western Australia.

Meagher, J.W. and Rooney, D.R. 1966. The effect of crop rotations in the Victorian Wimmera on the cereal cyst nematode (*Heterodera avenae*), nitrogen fertility and wheat yield. Australian Journal of Experimental Agriculture and Animal Husbandry 6: 425–431.

Millington, R.J. 1959. Establishment of wheat in relation to apparent density of the surface soil. Australian Journal of Agricultural Research 10: 487–494.

Millington, R.J. 1961. Relations between yield of wheat, soil factors and rainfall. Australian Journal of Agricultural Research 12: 397–408.

Morrison, D.A., Kingwell, R.S., Pannell, D.J. and Ewing, M.A. 1986. A mathematical programming model of a crop-livestock farm system. Agricultural Systems 20: 243–268.
Puckridge, D.W. and French, R.J. 1983. The annual legume pasture in cereal-ley farming systems of southern Australia: a review. Agriculture, Ecosystems and Management 9: 229–267.
Quigley, P.E. and Carter, E.D. 1985. The effect of cereal straw on productivity of annual medic pastures. Page 205 *in* Proceedings of the Third Australian Agronomy Conference, Hobart, Tasmania. Australian Society of Agronomy. Publisher?.
Reeves, T.G. 1987. Pastures in cropping systems. Pages 501–515 *in* Temperate Pastures: Their Production, Use and Management (Wheeler, J.L., Pearson, L.J. and Robards, G.E., eds). CSIRO, Melbourne, Australia.
Reeves, T.G. and Smith, I.S. 1975. Pasture management and cultural methods for the control of annual rye-grass (*Lolium rigidum*) in wheat. Australian Journal of Experimental Agriculture and Animal Husbandry 72: 527–530.
Robson, A.D. 1969. Soil factors affecting the distribution of annual *Medicago* species. Journal of the Australian Institute of Agricultural Science 35: 154–167.
Robson, A.D. 1983. Mineral nutrition. Pages 36–55 *in* Nitrogen Fixation, Vol. 3. Legumes (Broughton, W.J., ed.). Clarendon Press, Oxford, UK.
Robson, A.D. 1988. Nutrient requirements of pulses. Pages 869–881 *in* World Crops: Cool Season Food Legumes (Summerfield, R.J. ed.). Kluwar Academic Publishers, Dordrecht, The Netherlands.
Robson, A.D. and Abbott, L.K. 1987. Managing symbiotic associations between plants and microorganisms. Pages 191–203 *in* Temperate Pastures – Their Production, Use and Management (Wheeler, J.L., Pearson, C.J. and Robards, G.E., eds). CSIRO, Melbourne, Australia.
Robson, A.D. and Loneragan, J.F. 1970. Nodulation and growth of *medicago truncatula* on acid soils. II. Colonization of acid soils by *Rhizobium meliloti*. Australian Journal of Agricultural Research 21: 435–445.
Rossiter, R.C. 1966. Ecology of the Mediterranean annual-type pasture. Advances in Agronomy 18: 1–56.
Rovira, A.D., Smettem, K.R.J. and Lee, K.E. 1987. Effect of rotation and cultivation tillage on earthworms in a red-brown earth under wheat. Australian Journal of Agricultural Research 38: 829–834.
Rowland, I.C., Halse, N.J. and Fitzpatrick, E.N. 1984. The role of legumes in dryland agriculture. Pages 676–698 *in* Proceedings of the International Congress of Dryland Farming, Adelaide, 1980. Department of Agriculture, Adelaide, Australia.
Russell, J.S. 1981. Models of long term soil organic nitrogen change. Pages 222–232 *in* Simulation of Nitrogen Behaviour in Soil-Plant Systems. (Fissell, M.J. and Van Veen, J.U., eds). Pudoc, Wageningen, The Netherlands.
Sprent, J.I. 1976. Water deficits and nitrogen-fixing root nodules. Pages 291–316 *in* Water Deficits and Plant Growth. Vol. IV. Soil Water Measurements, Plant Responses and Breeding for Drought Resistance (Kozlowski, T.T., ed.). Academic Press, New York, USA.
Summers, R.N. 1987. The incidence and severity of non-wetting soils of the south coast of Western Australia, M.Sc. (Agric.) Thesis, The University of Western Australia, Perth, Australia.
Taylor, G.B. 1981. Effect of constant temperature treatments followed by fluctuating temperatures on the softening of hard seeds of *Trifolium subterraneum* L. Australian Journal of Plant Physiology 8: 547–558.
Taylor, G.B. 1984. Effect of burial on the softening of hard seeds of subterranean clover. Australian Journal of Agricultural Research 25: 201–210.
Taylor, G.B. 1985. Effect of tillage practices on the fate of hard seeds of subterranean clover in a ley farming system. Australian Journal of Experimental Agriculture 25: 568–573.
Taylor, G.B. and Ewing, M.A. 1988. Effect of depth of burial on the longevity of hard seeds of subterranean clover and annual medics. Australian Journal of Experimental Agriculture 28: 77–81.

Thomson, E.F., Rihawi, S. and Cocks, P.S. 1987. The survival of medic seed ingested by sheep. Pages 135–141 *in* Annual report for 1987, Pasture Forage and Livestock Program. ICARDA, Aleppo, Syria.
Thorn, C.W. and Perry, M.W. 1987. Effect of chemical removal of grasses from pasture leys on pasture and sheep production. Australian Journal of Experimental Agriculture 27: 349–357.
Tisdall, J.M. and Oades, J.M. 1980. The effect of crop rotation on aggregation in a red-brown earth. Australian Journal of Soil Research 18: 423–433.
Tuohey, C.L. and Robson, A.D. 1980. The effect of cropping after medic and non-medic pastures on total soil nitrogen and on the grain yield and nitrogen content of wheat. Australian Journal of Experimental Agriculture and Animal Husbandry 20: 220–228.
Tuohey, C.L., Robson, A.D. and Rooney, D.R. 1972. Moisture and nitrate conservation and responses to fallowing after medic ley in the Wimmera. Australian Journal of Experimental Agriculture and Animal Husbandry 12: 414–419.
Vallis, I., Henzell, E.F. and Evans, T.R. 1977. Uptake of soil nitrogen by legumes in mixed swards. Australian Journal of Agricultural Research 28: 413–425.
Venn, N.R. 1984. Control of annual grasses in pastures and legume crops with fluazitop-butyl. Australian Journal of Experimental Agriculture and Animal Husbandry 24: 612–616.
Vincent, J.M. 1974. Root nodule symbiosis with Rhizobium. Pages 265–341 *in* The Biology of Nitrogen Fixation. North Holland Research Monograph Frontiers of Biology, Vol. 33 (Quispel, A., ed.). North Holland Publishing Company, Amsterdam, The Netherlands.
Wells, G.J. 1969. Skeleton weed (*Chondrilla juncea*) in the Victorian Mallee. 1. Competition with legumes. Australian Journal of Experimental Agriculture and Animal Husbandry 9: 521–527.
White, D.H., Elliott, B.R., Sharkey, M.J. and Reeves, T.G. 1978. Efficiency of land-use systems involving crops and pastures. Journal of the Australian Institute of Agricultural Science 44: 21–27.
Willoughby, W.M. 1954. Some factors in grass-clover relationships. Australian Journal of Agricultural Research 5: 157–180.
Wolfe, E.C. 1985. Subterranean clover and annual medics – boundaries and common ground. *in* The Ecology and Agronomy of Annual Medics (Hochman, Z., ed.). New South Wales Department of Agriculture Technical Bulletin 32. Sydney, Australia.

Discussion

Buddenhagen:
Is it correct that your basement rocks are granitic and your land surfaces very ancient, thus contrasting greatly with the Mediterranean region?

Robson:
The soils of 'Mediterranean' Australia are very old, twice weathered, and very infertile compared to the soils of the Mediterranean, with very different levels of soil nitrogen.

Buddenhagen:
How much biomass remains of the pasture – roots and tops – to contribute organic elements to the following wheat crop?

Robson:
The pastures are continually grazed and often grazed heavily going into the cereal phase of the rotation, to get rid of grasses. Nitrogen fixed by legumes is thus recycled through animals as well as being contained in legume residues. Total production of roots may be 5–10 t/ha, with root:shoot ratio of 1:2.

Durutan:
Does soil depth have any effect on the performance of the legume pasture in rotation?

Robson:
Depth of ploughing is very important. Deep ploughing on deep soils will bury seed too deep for it to emerge.

Abd El Moneim:
Could you please elaborate on the effect of stubble on the hard-seededness of annual medics? How much is the recommended proportion of hard seeds for a new variety of medics to be acceptable in a ley-farming system?

Robson:
Stubble may reduce diurnal temperature fluctuations and thus decrease the rate of softening of hard seeds, leading to low plant populations. There is no single answer to the second question because it will vary with:
a) the rotation chosen (e.g. how many crops between pasture years in the rotation)
b) climate (e.g. frequency of 'false breaks', drought years when no seed is set, etc)
c) cultivation procedures used (e.g. amount of seed too deep to emerge). We need to follow the dynamics of seed numbers in each environment to be sure that there is an adequate plant population for maximum winter production.

Halila:
Is there any data comparing soil moisture conservation after medic and after fallow?

Robson:
Fallowing (land kept free of weeds) for either 6 or 9 months after medic pastures increased wheat yields primarily by increasing the amount of water conserved in the soil below 15 cm. Nitrate levels were increased after fallowing but we think if anything this may be a disadvantage in some situations because of 'haying-off'. Fallowing in South Australia only gave economic advantages when rainfall was less than 400 mm and soils had high water-holding capacity. There is very little crop-with-fallow in South Australia or Western Australia.

Abd El Moneim:
IN one of your figures you indicated that introducing annual medics reduced the population of cereal cyst nematodes on cereals. Is this because medics are not a good host for these nematodes? Have you a problem with root-know nematodes on medics, as we do?

Robson:
Yes, medics are not a good host for cyst nematodes. I am not aware of any problem with root-knot nematodes of medics in Australia.

Capper:
What is the effect of a one-year pasture break in a longer rotation where fields have kicked over into take-all decline?

Robson:
I am not aware of any hard data but would think that a one-year pasture break may reverse take-all decline.

Capper:
How specific are Rhizobium isolates to lucerne cv. CUF 101?

Robson:
In general lucerne has a wide range of ability to nodulate with a wide range of rhizobial strains.

PART IV

THE ROLE OF SOCIO-ECONOMICS AND EXTENSION IN PROMOTING LEGUMES IN FARMING SYSTEMS

Annual Cropping under Dryland Conditions in Turkey: A Case Study

N. DURUTAN, K. MEYVECI, M. KARACA, M. AVCI and H. EYUBOGLU

Field Crops Improvement Center, P.K. 226, Ulus-Ankara, Turkey

Abstract. After a brief description of the past and present agricultural situation in Turkey this paper gives an account of two development projects aimed at the utilization of fallow areas: the Corum-Cankiri Rural Development Project, begun in the late 1970s, and the Utilization of Fallow Areas Project, started in 1982 and now into its second phase. Its research is run by four Agricultural Research Institutes, and focuses on three main issues; crop improvement, crop rotations and crop management. It is backed up by a strong and effective extension program.

Introduction

Turkey is a country with a population of over 50 million in an area of 780 000 km^2. The country is characterized by extreme geo-climatic diversity which permits a wide range of crops to be grown under both rainfed and irrigated conditions. Some regions are even suitable for multiple cropping. Since the establishment of the Turkish Republic there has been a substantial development in the agricultural sector. Beginning in 1923, agricultural land has increased from 11.7 to 28.5 million hectares. This reflects an expansion in total arable land of about 2.5 times (Turkish MAFR 1987). The population of the country has increased approximately five-fold, from 10.5 million in the early 1920s to 50 million. Furthermore, productivity per hectare has gone up 8 to 10 times, depending on the type of agricultural product (Turkish MAFR 1987). All these developments have made Turkey one of the few countries in the world which are self-sufficient in food.

Wheat and barley are grown in almost every region in Turkey. Although production areas can be classified into 9 zones, actually there are 3 major wheat environments: winter wheat areas, spring wheat areas, and facultative wheat areas (Fig. 1). Most of the wheats produced in the country are winter wheats, primarily produced on the Central Anatolian Plateau and on its eastern and south-eastern extensions, and in Thrace. In these regions fallow-wheat and/or fallow-barley have been the primary cropping sequences for centuries (Fig. 2). Fallow has been practised not only because

A.E. Osman et al. (eds.), The Role of Legumes in the Farming Systems of the Mediterranean Areas, 239–255.

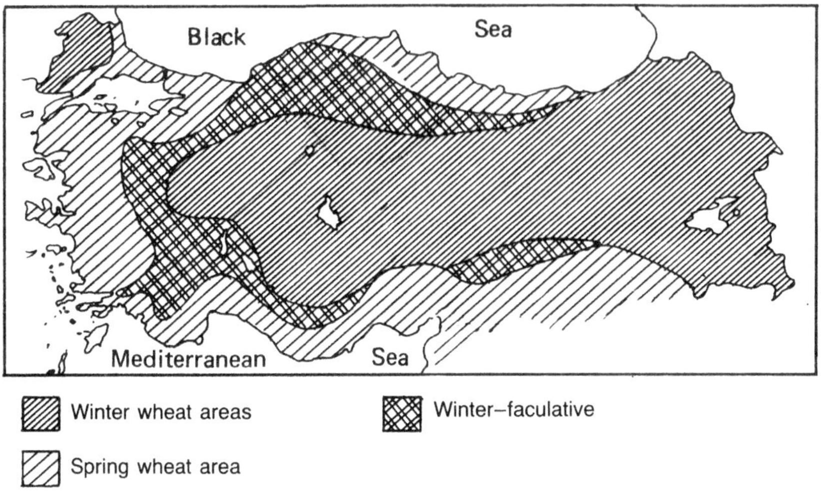

Fig. 1. Wheat-growing areas. (Durutan 1987).

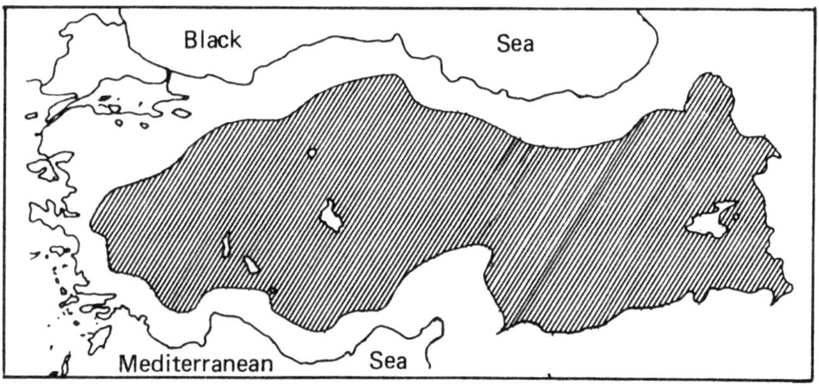

Fig. 2. Areas with fallow phase in the cropping sequence. (shaded). (Durutan 1986).

of the limited and variable annual precipitation but also because of long tradition. Technological development and modernization in agriculture have led to a remarkable increase in wheat production over the last 60 years (Fig. 3).

It can be seen that the production increase up to the late 1960s was mainly due to the expansion in area. Figure 3 shows that the acreage curve lost its sharpness while the production curve continued to rise with the yield curve. However, the increase in yield was the main reason for the production increase over the last 20 years.

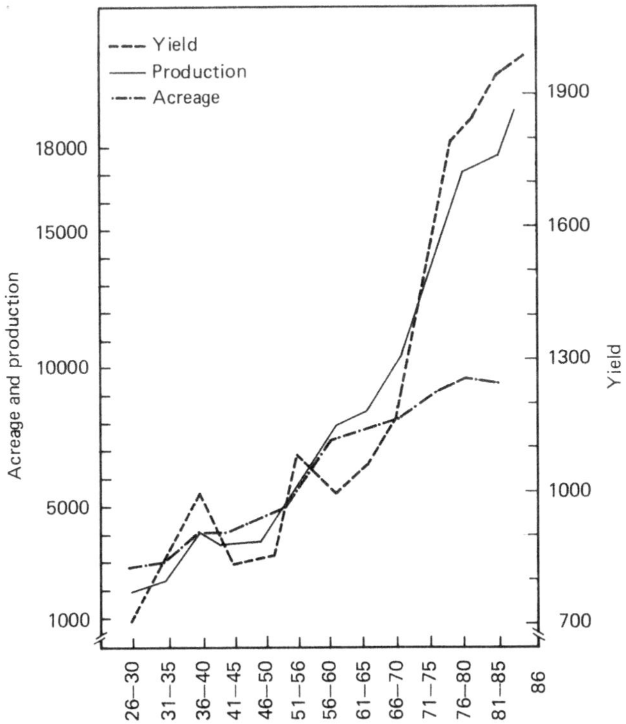

Fig. 3. Wheat acreage (1000 ha), yield (kg/ha), and production (1000 t), 1926–86 (Durutan 1987).

As mentioned above, fallow-wheat has been the primary cropping sequence for centuries. This means that in many regions the land use potential has not been fulfilled.

Utilization of Fallow Areas

The first attempt to utilize the fallow areas was made in the late 1970s by the Corum-Cankiri Rural Development Project in the north transitional area (Fig. 4). The total cost of the project was 162 million USD. The Government of Turkey provided 54% of the total and the rest was contributed by the World Bank. It was an integrated project aimed at the socio-economic development of the rural areas (Sahin 1981).

In both provinces most of the precipitation occurs in spring and winter (Fig. 5). The average long-term precipitation for Corum is 444 mm, and for Cankiri 390 mm. Average monthly temperatures are above O°C all through the year. The field crop area in Cankiri is 33% and in Corum 46% of the total cultivated land. The majority of the soils are heavy textured and mostly poor in phosphorus and organic matter.

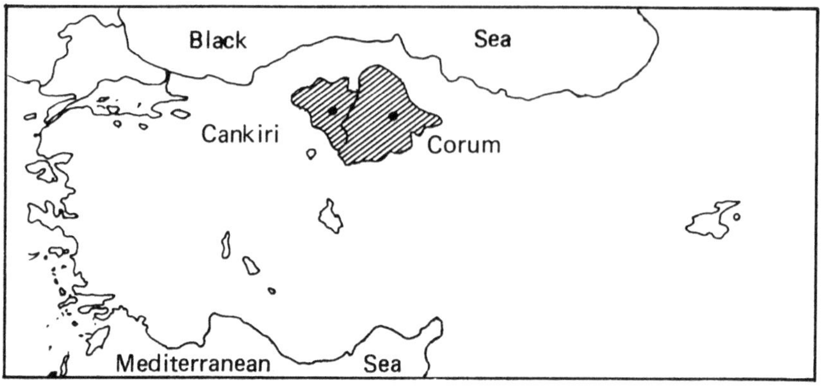

Fig. 4. The area of the Corum-Cankiri Rural Development Project.

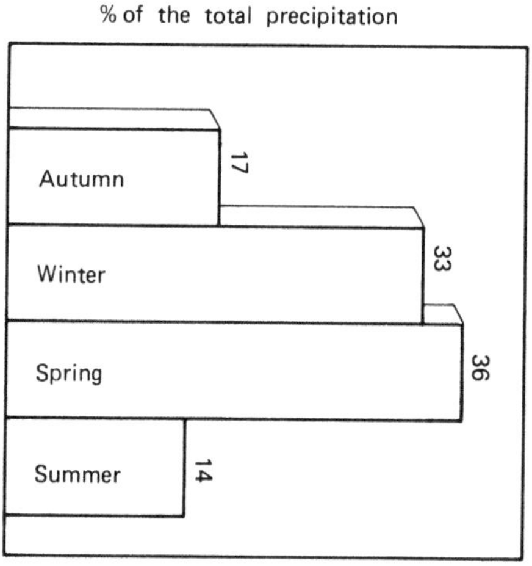

Fig. 5. Distribution of annual precipitation in the Corum-Cankiri area (Okcu 1981; Sahin 1981).

Before the implementation of the Rural Development Project fallow-wheat or fallow-barley were the main cropping sequences. In livestock-oriented farms a spring vetch for grain-wheat rotation was the most common system (Okcu 1981). Food legumes like chickpea and lentil were produced on many farms, however, for domestic use. Seed sizes were small and the yields were low.

The project provided large-seeded red and green cotyledon lentil seed and distributed it to selected farmers. Under the supervision of the project specialists the first promising yields were obtained. In Cankiri in 1977 yields were 1.2 t/ha from the red cotyledon type and 0.9 t/ha from the green (Okcu 1981).

Demonstrations were conducted with chickpeas too, and yields of about 0.9–1.1 t/ha encouraged the farmers to include these food legumes in rotations with wheat. The prices also influenced the farmers' decision (Okcu 1981).

In 1979 the first Hungarian vetch (*V. pannonica*) demonstration indicated that winter feed legumes were more advantageous in the rotation system than spring types.

The project required the use of inputs and the reduction of fallow areas, so more credits were made available for legume production. This played a vital role in the adoption process. Farmers were greatly attracted to legume production as an alternative to fallow and within five years a legume-wheat rotation system was largely adopted: legumes were grown instead of leaving the land fallow, so legume acreage increased as fallow areas decreased (Figs 6, 7 and 8).

The successful results obtained by this project served as a base for the next extension and research project, 'Utilization of Fallow Areas' (NAD). This is directed towards the utilization of fallow areas by using alternative crops. The activities are grouped under two disciplines:

1. Extension: to transfer the previous information to the farmers; to provide credits in cooperation with the Agricultural Bank; to provide necessary inputs; to evaluate the results.
2. Research: to identify the problems and conduct research to determine the necessary methods and measures for solving them.

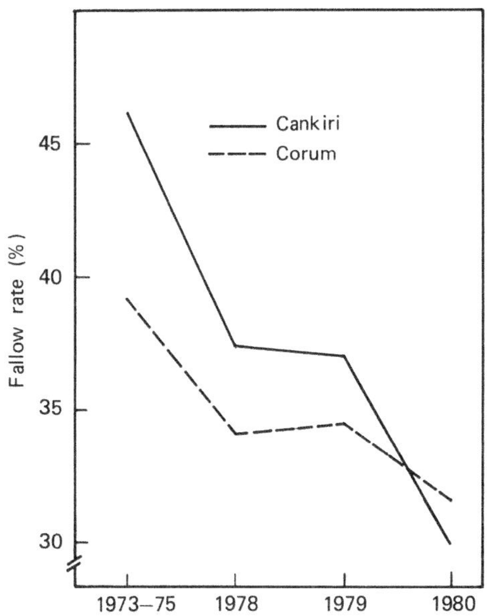

Fig. 6. The reduction of fallow areas in Corum-Cankiri as a result of the project (Okcu 1981; Sahin 1981).

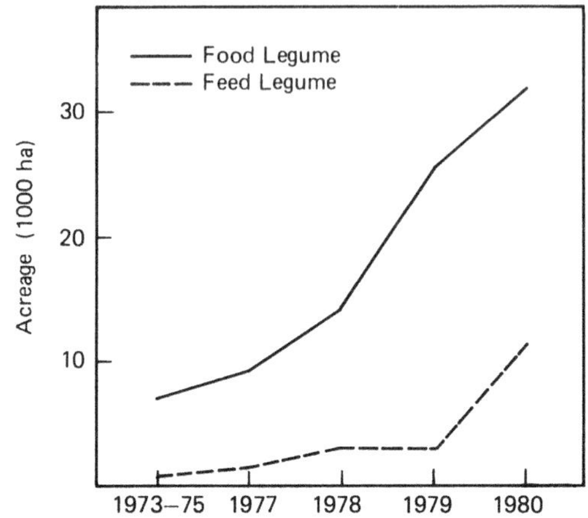

Fig. 7. Increase in legume acreage (1000 ha) in Corum in project years (Sahin 1981).

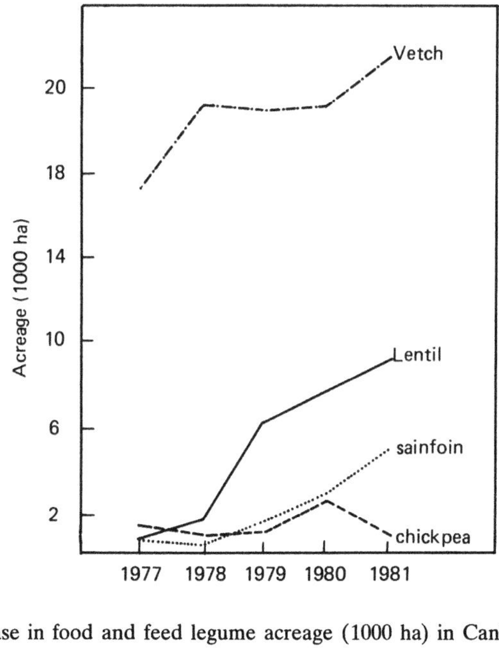

Fig. 8. Increase in food and feed legume acreage (1000 ha) in Cankiri (Okcu 1981).

At the beginning of the NAD Project there were two questions to be answered:

- Under which ecological conditions could fallow be safely eliminated?
- What are the most suitable and/or profitable crop rotation systems?

Since annual precipitation and/or soil moisture is the main reason for fallow practice, water should be the most important criterion for determining the fallow boundaries. From that standpoint, the precipitation data, previous research data and farmer practice were all taken into consideration and certain areas were grouped together as the pilot areas. The first pilot area consisted of 14 provinces and the second 12 provinces.

It was decided that in the first step 1.4 million hectares, and in the second step 1.7 million hectares of fallow areas could be utilized (Fig. 9). The first phase of the project was started in 1982 and completed at the end of 1986. The second phase will cover the period 1987–91 (Eskisehir 1987).

Based on previous research data, experience from the Corum-Cankiri Rural Development Project, and observations in the region, lentil, chickpea and vetch were recommended for flat land, while farmers are being encouraged to grow sainfoin (*Onobrychis sativa*) in the areas with a slope of more than 8%.

In order to reach the target some incentives are given to producers. The government has adopted support prices for these crops and farmers are entitled to obtain seed purchase and fertilizer credits. These incentives and measures have hastened the adoption of the new practice.

Supported by a strong extension program, within the period 1982–1986 the fallow areas decreased significantly (Table 1). The reduction of fallow areas has become very popular among the farmers, and fallow practice has been gradually abandoned not only in the project areas but in many other locations as well. According to the 1986 agricultural statistics, compared to 1979 the fallow areas around the pilot area decreased by 22%, and throughout the country they decreased by 31% (Table 1). The prices and guaranteed purchase by the government agencies hastened the process. Many

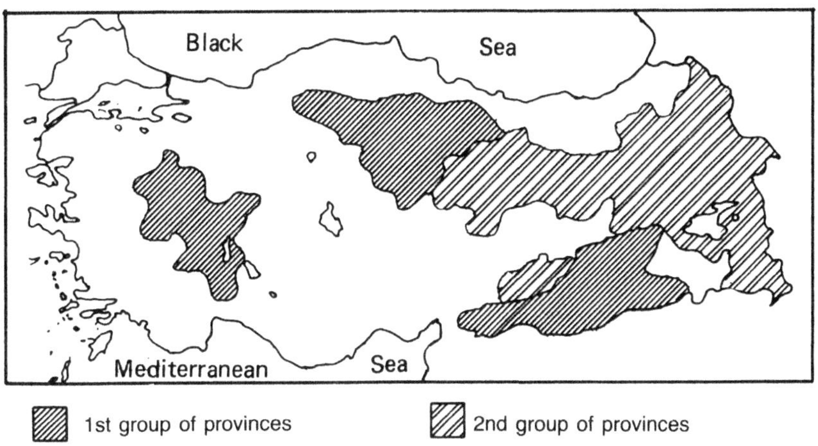

Fig. 9. The project areas where fallow could be utilized.

Table 1. Decrease in fallow (1000 ha) in the project areas; adjacent to the project area; and throughout Turkey.

The project area			Adjacent to the project area			Turkey		
1982	*1986*	decrease (%)	*1979*	*1986*	decrease (%)	*1979*	*1986*	decrease (%)
1434	939	35	6169	4832	22	8388	5771	31

Source: Devlet (1950–86).

farmers, convinced of the benefits of the new practices, have begun to crop annually.

Research Activities

The NAD Project requires the contribution of the research institutes in solving the problems of annual cropping. Four institutes (Ankara, Eskisehir, Diyarbakir and Erzurum) are responsible for related research. The main research can be grouped into 3 sections:

- Crop improvement
- Crop rotation systems
- Crop management practices

Crop Improvement Research

Unavailability of disease-tolerant, high-yielding winter-type legume varieties has been the major limitation to productivity in chickpea and lentil crops.

The lentil breeding program is being carried out with the following objectives:

- high yielding ability
- resistance to cold and drought
- simultaneous maturity
- large seed size
- suitability for mechanized harvesting

Until now, three winter, and four spring lentil varieties have been registered; three of them are green-seeded and four are the red cotyledonous type. Seeds are small-to-intermediate in size. Recently two new winter lentil lines were offered for registration. These are the first large-seeded winter cultivars developed by the program.

The chickpea breeding program is oriented according to the following objectives:

- resistance to Ascochyta blight
- cold and drought tolerance
- large seed size

- suitability for mechanized harvesting
- satisfactory cooking quality
- high protein content

Until now, two chickpea cultivars have been registered: one of these varieties is susceptible to *Ascochyta* blight, but the other is tolerant.

The cultivars developed by the national project are currently being used in crop studies and crop management research.

Rotation Research

Rotation studies were initiated in the 1930s in Eskisehir Dryland Farming Station and continued for 20 years in two different periods (Kalayci 1987). The results are summarized below:

1. In the 2-year rotation system the best yields were obtained from fallow-barley and fallow-wheat. The long-term wheat yield data indicated that the legumes were the best crops in terms of the effect on the succeeding wheat yield level. Among the legumes the giant vetch (*Vicia narbonensis*) was found to the the most promising crop since it is planted in winter and cut as a green forage at the beginning of the summer. The yields of the spring legumes were found to be less stable than those of winter types since they depend upon the spring rains.
2. The fallow-wheat-giant vetch-wheat cropping sequence was the best 4-year rotation system.
3. Three-year sainfoin (*Onobrychis sativa*) followed by wheat for 3 years was not a good system. The negative effect of the perennial crop plus the continuous small grain lowered the final wheat yield drastically.
4. The data indicated that the effect of the previous crop becomes less pronounced as the annual precipitation rises above 400 mm. Where there is 500 mm of precipitation the differences between the various rotation systems are almost eliminated (Fig. 10).

Under limited rainfall conditions cropping depends upon the stored moisture in the soil. Cutting the legumes early in the spring provides more moisture to the succeeding wheat than growing crops for grain. Another advantage is the availability of the land for better seedbed preparation for wheat. Kalayci (1987) reported that because of these positive points a feed legume-wheat rotation yields better than grain legume-wheat (Table 2). In Kalayci's trial, the yield differences among the treatments were insignificant in wet years and vice versa.

Another study was conducted in the 1960s to see the effect of the utilization of the feed legume on the succeeding wheat yield (Table 3).

Table 3 indicates that the yield of wheat after feed legume for seed is lower than after the legume cut for green forage, particularly in dry years like 1967.

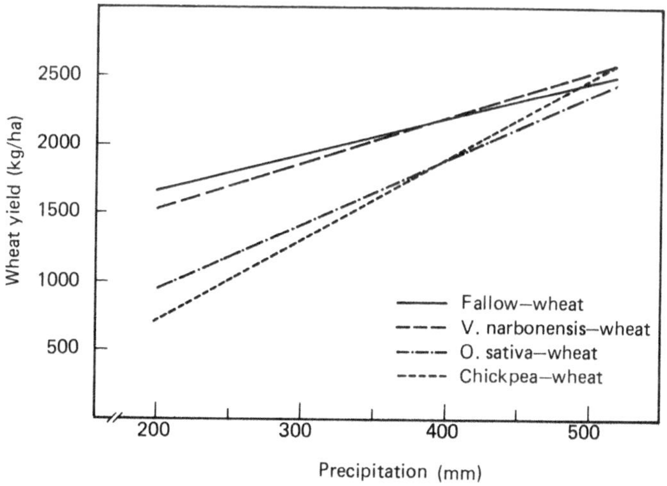

Fig. 10. Relationship of yield (kg/ha) × precipitation (mm) on rotation system (Kalayci 1987).

Table 2. Effect of different crop rotation systems on wheat yield in relation to annual precipitation.

	Wheat yield (kg/ha)				
Crop Sequence	1965	1966	1967	1968	Mean
Fallow-wheat	1680	2090	1440	1350	1640
V.narbonensis-wheat	1570	1740	1100	1260	1420
V.villosa-wheat	1430	1860	1240	1240	1440
P.arvense-wheat	1430	1890	990	1320	1410
Precipitation (mm)	355	445	239	427	

Table 3. Effect of utilization of the feed legume on wheat yield.

	Wheat Yield (kg/ha)				
Utilization	1965	1966	1967	1968	Mean
Cut for green forage	1570	2190	1250	1400	1600
Seed production	1310	1540	700	1190	1190
Precipitation (mm)	355	445	239	427	

Source: Kalayci (1987).

With the initiation of the NAD Project special emphasis was placed on crop rotation research. Currently, various experiments are being conducted to determine the optimum fallow frequency and the most economical crop rotation system.

Since sufficient data have already been developed to get an overall idea the data obtained from the 2-year rotation experiment will be given below (Table 4). The differences among the treatments were found to be signifi-

Table 4. Soil moisture(mm), nitrogen (kg/ha), and yield (kg/ha) data obtained in 2-year rotation experiment, Haymana, 1982–86.

Treatments	Soil moisture[1]	NH_4	Nitrogen No_3	Total inorganic	Yield,[2] 1st Crop	Yield,[2] wheat
Sunflower-wheat	201	79	27	106	900	2780
Winter lentil-wheat	238	106	30	136	1170	3140
Spring lentil-wheat	209	107	30	137	940	2930
Hungarian vetch-wheat	238	110	27	137	1850	3230
Safflower-wheat	192	60	5	65	1010	2340
Barley-wheat	195	54	10	64	1790	1770
Cumin-wheat	244	93	31	124	580	2860
Chickpea-wheat	204	110	20	130	1430	2910
Fallow-wheat	281	101	39	140	–	3230

[1] Average of 2 years (1985–86).
[2] Average of 5 years (1982–86).

Source: Meyveci (1988).

cant. Fallow was found to be significantly better than other treatments in terms of moisture storage. The data indicated that the crops that complete their growth before the onset of the hot dry period leave the highest amount of moisture for wheat. In other words, winter legumes gain an advantage over the summer crops in the rotation system. Among the summer crops legumes performed better than sunflower, safflower and barley. Results obtained with barley indicated that continuous small grain is not a good cropping sequence in terms of moisture accumulation.

The nitrogen status of the rotation system in the 0–120 cm soil profile was determined as ammonium and nitrate nitrogen. Average values showed that ammonium nitrogen provided by legumes was at least as much as by fallow. The differences among the treatments were significant and legumes and fallow were in the same group. Two-thirds of the total ammonium nitrogen was accumulated at the upper layers of the profile. It decreased gradually with increasing depth.

The amount of total nitrate nitrogen also differed significantly among the treatments (Table 4). Nitrate nitrogen was found to be significantly higher in fallow plots, with the legume plots, except chickpea, next. The lowest amount of nitrate nitrogen was determined in safflower and barley plots. In fallow plots more than half of the total nitrate nitrogen was accumulated at 0–30 cm depth.

The data also indicated that fallow and legumes provided the highest amount of total inorganic nitrogen.

Yields of wheat after legumes, especially winter-sown, were nearly as high as after fallow.

Crop Management Research

Numerous experiments are being carried out to determine the proper management package. This is important not only for the rotating crop itself but also for the wheat.

Proper seedbed preparation system, seeding date, method and rate of fertilizer application, and weed control are also being studied (TARM 1981–87).

As is the case for many crops, stand establishment is very important for legumes. Seedbed preparation should get priority in the management package since stand establishment largely depends on the seedbed properties. The Field Crop Improvement Center (TARM) is carrying out soil tillage experiments in winter vetch-wheat and spring lentil-wheat rotation systems (TARM 1981–87).

In the 2-year rotation experiment (see Table 4) various crops are being rotated with wheat, such as Hungarian vetch, barley, lentil (as winter crops); and chickpea, lentil, sunflower, and safflower (as spring crops). These systems are being compared to the fallow-wheat sequence.

The cropping periods within 26 months (fallow year + wheat year) are given in Fig. 11.

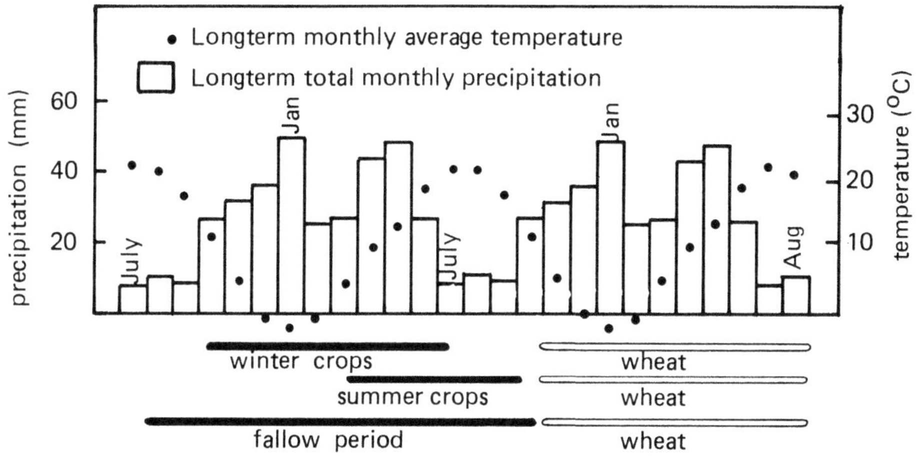

Fig. 11. Cropping periods over 26 months.

The 2-year average of soil moisture and nitrogen data collected at wheat seeding time from the 0–120 cm soil profile is given in Table 4.

TARM placed special emphasis on weed control measures since yield reduction due to weeds is significant, particularly for the winter-planted crops. Various experiments are being conducted to determine the most effective cultural control method as well as chemical methods. Since an integrated weed control approach has been adopted, various treatments in every component of the management package are being tested to determine their effectiveness on weed control.

The data obtained from soil tillage experiments carried out in the winter vetch-wheat system are given in Figs 12 and 13. Data of three cycles are

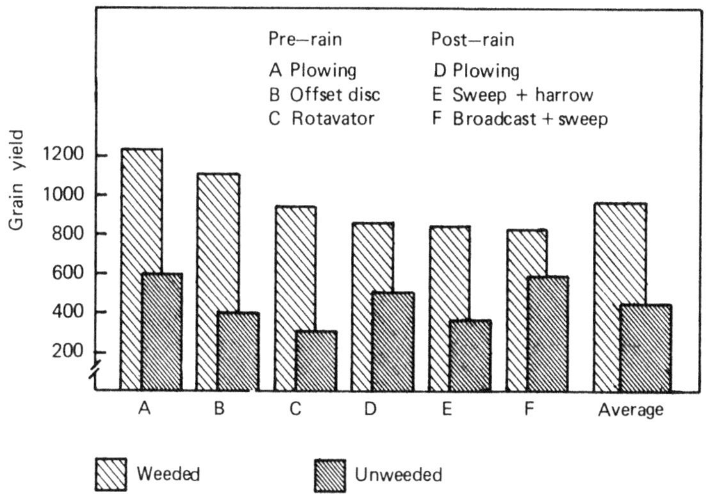

Fig. 12. The effect of soil tillage on grain yield (kg/ha) of winter vetch, Haymana, 1986/87 (TARM 1981–87).

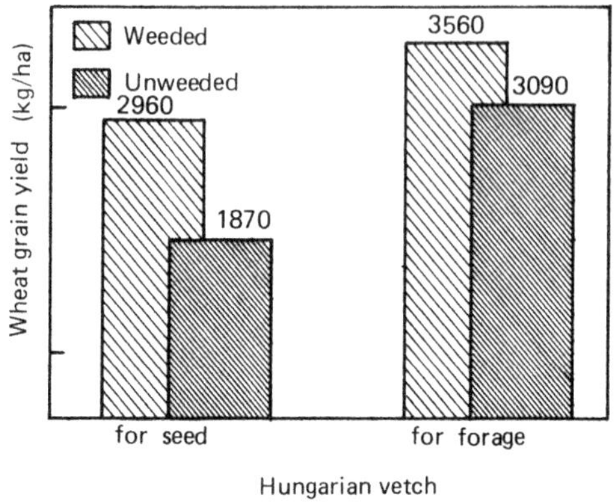

Fig. 13. The effects of seed and forage production of winter vetch on wheat yield (kg/ha), Haymana, 1984–87 (TARM 1981–87).

being collected on soil moisture, grain and forage yield of vetch, and grain yield of wheat under weeded and unweeded conditions. Leaving the crop unweeded reduced the grain yield significantly. Yield decreases ranged from 29% to 68%. Grain production negatively affected the yield of the next crop. The wheat yield decreased 17% compared to forage production of the previous crop. Weeds in grain vetch plots had a greater negative effect on wheat yields than weeds in forage vetch plots.

For the spring legume(lentil)-wheat system another soil tillage experiment is being conducted in the Central Anatolian Plateau. Data are being collected on soil moisture and some physical properties of soil, and on lentil and wheat yield (Figs 14 and 15).

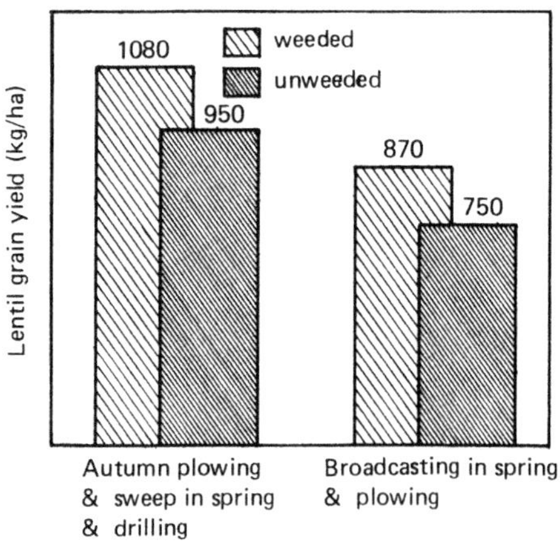

Fig. 14. The effects of soil tillage and sowing method on lentil yield (kg/ha), Haymana, 1984–87 (TARM 1981–87).

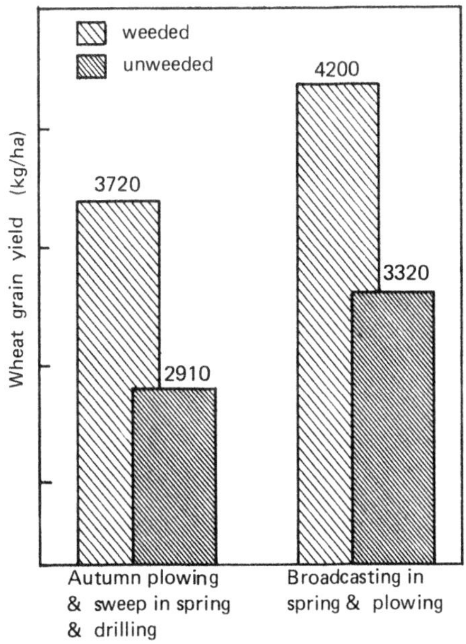

Fig. 15. The effects of seedbed preparation methods and weed control on wheat grain yield (kg/ha), 1985–87.

'Autumn plowing + sweep in spring + drilling' yielded 24% more than 'broadcasting and covering the seed with mouldboard' under clean conditions. In comparison to a weeded environment the average yield decrease was 13% under unweeded conditions in both tillage systems (Fig. 14).

The effect of the seedbed preparation system on the following wheat is given in Fig. 15. The seedbed preparation and seeding methods for spring lentil significantly affected the wheat yields in the second year. Seeding the lentil by 'broadcasting + plowing' yielded 13% more than 'autumn plowing + sweep in spring + drilling' due to easier seedbed preparation for wheat in autumn. Leaving the lentil unweeded reduced the wheat yield 21% on average.

In both experiments it was observed that the negative effect of soil tillage systems and weed competition accumulated over the years.

Within a few years the management package will be completed and transferred to the farmers. Then it will be possible to observe the impact of research on the national yield level.

References

Devlet Istatistik Enstitusu 1950–86 Turkiye Istatistik Yillari.

Durutan, N., Pala, M., Karaga, M., Yesilsoy, M.S. 1986. Soil Management, Water Conservation and Crop Production in the Dryland Areas of Turkey. International Wheat Conference, May 2–5 1986. Rabat, Morocco.

Durutan, N., Yilmaz, B. 1987. Small grains and food legumes production improvement in Turkey. Pages 66–80 *in* Winter Cereals and Food legumes in Mountainous Areas. Proceedings of Symposium, July 1987, Ankara, Turkey (Srivastava, J.P., Saxena, M.C., Varma, S. and Tahir, M. eds). ICARDA-136 En, Aleppo, Syria.High Altitude Symposium, 6–10 July. Ankara, Turkey.

Eskisehir Zirai Arastirma Enstitusu. 1987 NAD Projesi 1982–86 Donemi Toplu Degerlendirme Raporu.

Kalayci, M. 1987. Eskisehir Zirai Arastirma Enstitusu tarafindan bugune kadar yapilan nadas alanlarini azaltmaya yonelik calismalar. Kuru Tarim Bolgelerinde Nadas Alanlarindan Yararlanma Simpozyumu, 28–30 Eylul, 1981, Ankara. Tubitak Yayinlari No.593.

Meyveci, K. 1988. Orta Anadolu Bolgesi Kosullarinda Ikili Ekim Nobeti Sisteminde Toprakta Nem ve Inorganik Azot Formlarinin Belirlenmesi. A.U.Z.F. Doktora Tezi.

Okcu, S., Dogru, S. 1981. Kuru Tarimda Nadasi Azaltici Calismalar. Tubitak Kuru Tarim Bolgelerinde Nadas Alanlarindan yararlarma Simpozyumu. 28–30 Eylul, 1981, Ankara, Turkey.

Sahin, H. 1981. Corum-Cankiri Kirsal Kalkirma Projesi 'Nadas Alanlarinin Azaltilmasi Calismalari'. Tubitak Kuru Tarim Bolgelerinde Nadas Alanlarindan Yararlanma Simpozyumu. 28–30 Eylul 1981. Ankara, Turkey.

TARM, 1981–87. Yetistirme Teknigi Yillik Gelisme Raporlari.

Turkish Ministry of Agriculture, Forestry and Rural Affairs, 1987. Agriculture in Turkey.

Discussion

Jones:
We have heard nothing about pasture work in Turkey. Is this because pastures are not considered important, for reasons of altitude (and therefore climate), because of the nature of the farming systems, or because of government policies?

Durutan:
Unfortunately not much pasture work has been carried out in Turkey. We collected some information but couldn't transfer it to the farmer. I believe we didn't try hard enough to overcome the socio-economic barriers.

Osman:
How were weeds controlled in Hungarian vetch?

Durutan:
In the farmer situation there was no weed control, so the quantity of hay was large but the quality was not good. On our plots we used hand-weeding and cultural practices: chemical methods would not be so relevant to the farmer situation.

Osman:
What was the main reason for the sharp reduction in the fallow area?

Durutan:
The main reason was the buying guarantee by the government buying agencies. The price was good too, and after the initial period the possibility of export also contributed.

Solh:
With respect to the response to N fertilization of the three legumes, how does their effective nodulation compare?

Durutan:
Under our conditions, chickpea is best, then Hungarian vetch, but lentil is not very good.

Ibrahim:
What is the ratio in the acreage of winter and spring lentil?

Durutan:
We only grow winter lentil in South-eastern Anatolia, on about 60% of the total acreage.

Robson:
When comparing winter- and spring-sown legumes, we must consider that there may be delayed nodulation at lower temperatures. As well as selecting cultivars of plants well-adapted to cold it may be valuable to select Rhizobia able to nodulate well in winter conditions.

Solh:
Do you have data on the effect of unweeded and weeded spring chickpea on the yields of the following wheat crop?

Durutan:
We have no data, but we have observed that the weed problem with chickpea is much less than with any other spring legume. I don't think that it has more than a 10% negative effect on the succeeding wheat crop.

Osman:
There has been a common complaint in the presentations of the last two days about the inadequate extension service in some countries. You are the first to speak positively about a strong service in Turkey. I believe this is an important development. Could you please elaborate?

Durutan:
Research and extension work together; we hold monthly workshops where the extension people bring the problems of farmers, and the research people try to find the solution. If there is no ready solution, the problem is included in the research program. Every year, for the Utilization of Fallow Project, research and extension workers come together and discuss the findings of research, and the bottlenecks and problems in the implementation of techniques. At the Field Crops Improvement Centre we are running many training programs. Extension people tend to come straight from university, with no specialised training in extension, so they often feel very insecure. By giving them training, and by working together with them, we are trying to increase their knowledge, and especially their confidence.

Nassib:
Would you elaborate a little on how on-farm trials on food legumes are conducted?

Durutan:
Our basic research is being conducted at the research farm. The most promising two or three treatments are carried to farmers' fields, either replicated, or on larger plots with no replications, depending on the type of experiment. Usually the plot sizes are around 1000–2000 m^2. These on-farm trials are conducted for two or three years. They are planned together with the extension staff and conducted cooperatively. Modifications are made where necessary, depending on the results and farmer attitudes.

The Extension of the Ley Farming System in South Australia: A Case Study

G.D. WEBBER

South Australia Department of Agriculture, Adelaide, Australia

Abstract. Agricultural development in South Australia has gone through three phases: (i) land clearance and continuous cereal cropping by the early settlers, leading to a drop in soil fertility and cereal yields; (ii) the introduction of new wheat varieties and the increased use of superphosphate, and cultivated fallowing in rotation with cereals, leading to damaged soil structure and severe erosion; and (iii) from about 1940, the development of the ley farming system, after recognition of the benefits of incorporating annual self-regenerating pasture legumes into a crop rotation. The author lists the key features of this integrated crop-livestock system, points to the results attributed to it, i.e. increased grain yields, animal numbers and wool and meat production, and describes the technical problems and extension methods affecting its adoption by Australian farmers. Lessons from this experience are related to agriculture in Mediterranean areas where ley farming could be relevant. The need for careful modification of the system and expert extension is stressed, in view of its complexity and the many constraints to adoption in these areas.

Introduction

Agricultural production in the semi-arid area of South Australia (annual rainfall 250–550 mm) has largely been based on growing field crops, mainly cereals (wheat, barley and oats), in rotation with annual legume pastures, mainly *Medicago* spp. This farming system is called ley farming, and integrates cereal and livestock production. The legume pastures with their nitrogen-fixing properties increase soil fertility for cereal crops and supply high quality feed for large numbers of livestock.

South Australia has a Mediterranean-type climate characterised by a dry summer and short winter rainfall season of 5–6 months which allows the growth of short season crops and legume pastures. There is a large variation in annual rainfall, and droughts are common.

Climate and Land Use Zones

In regions with Mediterranean climates rainfall mainly determines land use, and the cereal producing zones are placed between the arid zone (where

A.E. Osman et al. (eds.), The Role of Legumes in the Farming Systems of the Mediterranean Areas, 257–272.

agricultural production is mainly based on extensive low density grazing of the native shrub steppe and where cropping is only normally possible with irrigation), and the humid zones, where lack of moisture is not a restraint.

In South Australia the three major zones of agricultural land utilization are:

1. The dry inland pastoral area, with less than 250mm annual rainfall, has a low-intensity grazing system based on a delicate balance of utilization and conservation of native vegetation. The main enterprise is sheep-grazing, supplemented with some cattle. Stocking-rates are controlled and vary from four to twenty sheep per square kilometre (one sheep per 5–25 ha).
2. The cereal zone is the intermediate rainfall area with 250–500 mm annual rainfall, where production is based on cereal crops in rotations with annual legume-based pastures grazed by livestock.
3. The high-rainfall zone with more than 500 mm annual rainfall, where the main activity is high-intensity grazing of sheep and cattle on sown and fertilized legume-based pastures, with lesser dependence on cropping.

The three zones are interdependent and integrated, which benefits the stability of each system.

Phases of Agricultural Development

Like many of the more arid dryland farming areas of the world, much of the southern Australian cereal zone has been through the exploitive stage. There have been three definable phases of development in the Australian wheatlands (Fig. 1).

Initially the early settlers, realising that cereals especially wheat, would

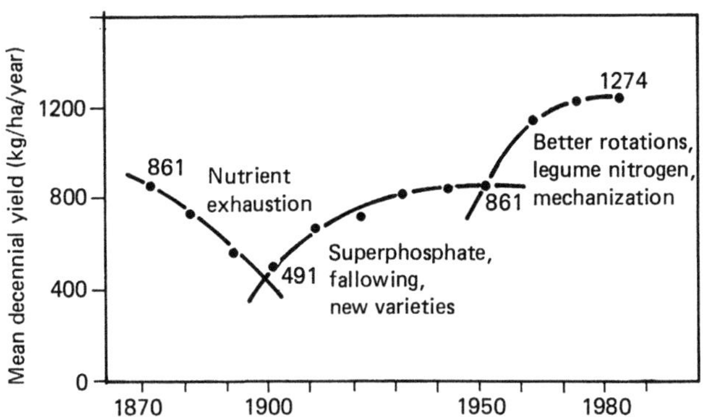

Each point indicates the unweighted mean of the annual yields in the preceding decade.

Fig. 1. Australian wheat yields, 1860–1980 (Donald 1981).

grow in the southern Australian cereal zone, cleared native vegetation, cropped continuously and the initial soil fertility soon fell, as did crop yields.

During the second phase (1900–1930), new wheat varieties were introduced, superphosphate came into universal use and a fallow-crop system was developed, supported by the belief that this helped to conserve moisture. However, this continual cultivation and cropping damaged soil structure. Land was left bare by fallowing and soon wind erosion led to severe dust storms in the summer. Winter rains gouged gutters into the unproductive soils of the hillsides and the limited depth of good top soil was severely reduced by sheet erosion.

We know now, of course, that the early farmers up to the late 1930's in the South Australian cereal zone exploited the limited resources by extensive cropping and fallow. As a result of this fallow-wheat system, much of the soil's natural fertility was exhausted, soil erosion was severe in many areas, soils were difficult to work because their structure had been broken down, and there was insufficient forage to feed increasing numbers of livestock.

The third phase of southern Australian agriculture followed recognition of the benefits of incorporating annual self-regenerating species of pasture legumes into a crop rotation. Areas sown to pastures increased rapidly from the mid to late 1930s to 1950. By 1967, more than 20 million hectares were sown to pasture in Australia (Macindoe 1975). A large part of this improved pasture area was sown in rotations with wheat.

Impact of the Ley Farming System in South Australia

The evolution of the ley farming system changed the farming outlook in the dry farming lands of southern Australia. Under this system, the bare fallow phase of the rotation in alternate years was largely replaced with a legume pasture ley. This farming system widely practiced in South Australia developed mainly in the period between 1940 and 1960.

The annual legume pastures grown between cereal crops are based on cultivars of *Medicago* species or *Trifolium subterraneum*. They form the basis of an integrated system of cereal and livestock production, which has resulted in vastly increased livestock production, higher soil nitrogen levels, a marked reduction in fallow and large increase in forage production. Under this system, the landholder owns his own livestock and each year has part of his farm under crop and part under pasture.

The net result has been that within the cereal production zone, crop production and yields per hectare have increased and considerably higher stock numbers, with increased wool and meat production, have been achieved.

The development of the ley farming system was quite rapid. There was a gradual start in the late 1930s and early 1940s but by 1950 annual legumes (mainly *Medicago* spp.) were sown or had volunteered over a wide area under adapted management techniques. They grew prolifically in many areas and were stimulated by phosphate applications. As well as supporting more sheep, the legume-based pastures rebuilt depleted organic matter reserves and lifted soil nitrogen levels. The increase in soil fertility was reflected in increased cereal crop yields and higher stocking capacities.

In the main cereal growing areas of the State, cereal yields rose by 50%, sheep numbers doubled, and wool yields more than tripled between the 1930s when annual legumes were first encouraged and the 1960s when their use was widespread (Table 1). The alkaline soil areas which grew *Medicago* spp. successfully, showed most rapid responses to increased soil fertility. A slower rate of development occurred on the neutral to slightly acid soils. The acid soil areas were found to be more suited to early cultivars of subterranean clover which had much lower levels of hard seed reserves and hence needed resowing after each period of cropping. However, in areas where subterranean clovers could be successfully sown, increases in soil fertility resulted in production gains at least equal to gains made in medic areas.

These increases are illustrated by data from the Turretfield Research Center (Fig. 2) that show (over 5-year average periods) that both wool and cereal yields doubled following the sowing of legume pastures (mainly subterranean clover).

An important part of the change to the ley farming system and its subsequent impact on agricultural production was the widespread adoption

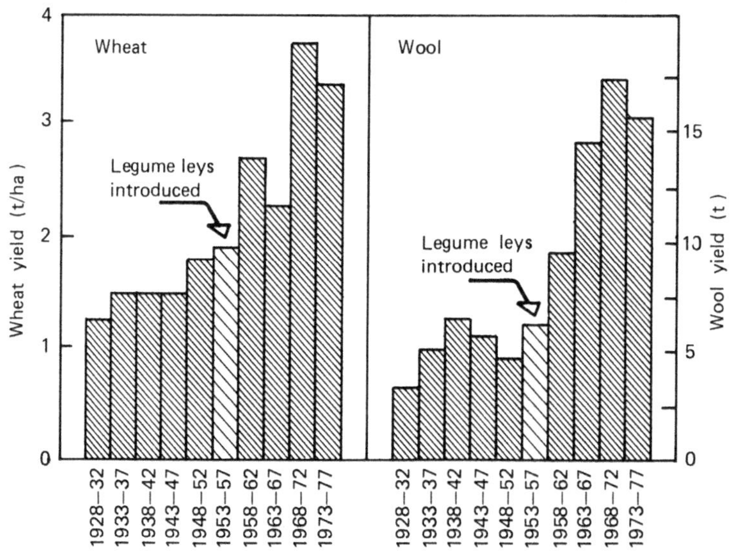

Fig. 2. Effect of legume leys on wheat (t/ha) and wool (1000 kg) production, 1928–77, Turretfield Research Centre (Webber et al. 1976).

Table 1. Cereal area (1000 ha) and production (1000 t), livestock numbers (1000) and wool production (1000 kg), South Australia – Average of Ten-Year Periods.

Period	Wheat		Barley		Total cereal production	Sheep and wool		Cattle
	Area	Prod.	Area	Prod.	Wheat, oats and barley	Numbers	Wool Prod.	Numbers
1931–40	1326	939	151	142.3	1125.5	8504	33731	331
1941–50	824	712	205	222.5	990.6	9226	42842	422
1951–60	633	779	468	579.4	1477.0	13658	70703	523
1961–70	1128	1272	503	537.5	1943.5	17 120	100 574	725

Source: Webber (1975).

of practices to control soil erosion from wind and water. These included a reduction in fallowing, modified tillage methods to prevent working soils to a very fine state, and the contour banking of large areas of undulating arable land that was susceptible to water erosion (some 220 thousand hectares of land in South Australia have been contour banked since 1946).

The Development and Extension of the Ley Farming System

The initial development of the *Medicago* based ley farming system was basically an extension exercise.

In the adoption stage, research and development inputs were limited and mainly related to the development of mechanical means of harvesting sufficient seed to distribute to farmers. As an extension exercise, it must be judged as a success, measured by the adoption of the system at farm level.

An analysis of the field application of the ley farming system, shows how it developed in three stages:

Stage 1

The first stage was based on *Medicago truncatula*, named barrel medic by Hutton (1934). It was presumably introduced from the Mediterranean region with the earliest botanical specimens collected near Mannum in South Australia in 1915.

Trumble (1939) working at the Waite Agricultural Research Institute first recognised the potential value of barrel medic in the early Burr medic (*Medicago polymorpha*), suggesting it could be as valuable on the alkaline soils as subterranean clover (*Trifolium subterraneum*) on the acid soils. Trumble introduced Hannaford (of the Agricultural Machinery and Seed Grading Co. of Alf Hannaford and Son) to barrel medic and they first collected seed by hand from an area south of Adelaide in 1934. From 1934 to 1936, Hannaford (who had the first barrel medic cultivar named in his honour) worked on more efficient methods of harvesting seed, devising a rake to remove top vegetative material, brooms to sweep pods into a wheel device and then on to an elevator to carry the pods into a winnower to remove foreign matter. Then pods could go into the thresher to separate the seed.

By 1937, A. Hannaford and Sons were harvesting commercial quantities of seed (then known as Commercial barrel medic), advertised its availability, and sold and distributed quantities to farmers around cereal farming areas of South Australia (Higgs, personal communication). The company produced information sheets and wrote articles for local newspapers on the value of this pasture legume. Field Officers of the Department of Agriculture at this time were also promoting barrel medic and distributing seed for demonstration purposes. Areas were sown at Roseworthy Agricultural College and then at Minnipa Research Center in 1944, where by 1947 work had commenced on harvesting medic seed (Day and Michelmore 1952). By 1946, large quantities (up to 50 t) of medic seed were being produced in South Australia and this had increased to 132 t by 1952.

From 1946 an economic boom lifted demand and price for livestock products, and with increasing amounts of superphosphate becoming available, there was a very rapid expansion of the area sown to annual legume pastures. In some cases these pastures volunteered, with pods being spread by sheep (pods clinging to the wool) and carried from field to field.

Stage 2

Medicago truncatula known then as 'Commercial' barrel medic had demonstrated that it grew best on the soils with high lime, particularly the rubbly limestone soils of the South Australian Wheat Belt. The main extension programs at this stage concentrated on

(i) the promotion of annual legumes to increase production of livestock feed and increase soil fertility;
(ii) the development of harvesting methods to produce sufficient seed;
(iii) the demonstration of the production potential of pastures fertilized with superphosphate;
(iv) the development and demonstration of methods of seeding – including the development of the small seeds box fitted to the back of cereal seeding machinery in order to sow seed at correct depth.

However, one problem that existed was the large areas of alkaline sandy and heavier textured soils on which barrel medic did not grow prolifically.

In fact, there had been a natural development of *M. polymorpha* (spiny burr medic) on these soils and *M. minima* (woolly burr medic) on the drier fringes with the addition of phosphate.

The next step forward was the introduction and development of a number of new cultivars, particularly *M. truncatula* cv Jemalong for the heavier textured soils, and *M. littoralis* cv Harbinger for the sandy soils.

Seed of *M. scutellata*, snail medic, had previously been imported from the United Kingdom and sown over a range of soil types. It grew prolifically on the heavy textured grey-black soils. However, because of its large pod and erect seeding habit, it was susceptible to overgrazing and also regenerated irregularly.

The field testing in South Australia of Jemalong (known earlier as Barrel Medic 173 in N.S.W.) started in 1951 and an intensive demonstration sowing program in the main cereal growing areas started in 1956. (Crawford 1962a).

There still existed a need to extend the range of medic cultivars in the lower rainfall areas below the areas of adaptation of Jemalong, and this was met by the development of Cyprus barrel medic (initial seed sowings in South Australia in 1958) for the loamier textured soils, and Harbinger strand medic (first sowing 1958) for the light sandy soils (Crawford 1962b,c).

By the early 1980s a number of cultivars had been developed. Their adaptation to a wide range of neutral to alkaline soils, and rainfall and growing season requirements (Fig. 3) was demonstrated in the range of farming systems.

The extension of these new cultivars, together with recommendations regarding seeding methods, pest control, fertilizer requirements, and grazing management techniques programs included demonstration of cultivars in field trial comparisons, Field Days and farmers' meetings, and the production of leaflets, Bulletins and Fact Sheets for farmers as well as a high level of programs on radio and television, articles in local newspapers.

Stage 3

The successful adoption and expansion of the ley farming system endured over almost 30 years. But this success was in the adoption of the ley farming system itself – the subsequent development of the system and incorporation of more sophisticated technology wasn't and isn't now as effective as might have been anticipated.

While South Australia has maintained its ability to carry a relatively stable livestock population, field crop yields are still low by world standards and by efficiency standards in relation to the potential yields per mm rainfall (French and Schultz 1984).

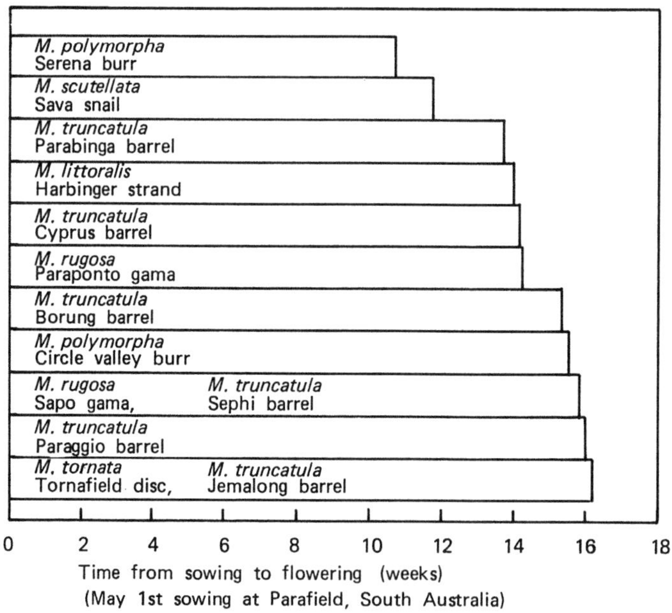

Fig. 3. Time from sowing to flowering of Australian cultivars of annual medics (Boyce and Crawford, personal communication).

In the 25 years 1945–1970, livestock products attracted prices which encouraged farmers to produce both cereals and livestock. Livestock production was therefore not only profitable but its incorporation into the system was also biologically sound in that the legume pastures assimilated nitrogen into the soil, improved soil structure and subsequently increased crop yields. But livestock prices fell in the late 1960s and early 1970s making cropping relatively more profitable and many farmers responded by increasing the intensity of cropping in some areas by up to 40%. Many farmers in the better rainfall areas now sow field crops on more than half their farm area; indeed some farmers in the better soil areas crop almost all their land area. This has placed great pressures on the pasture phase of the rotation and natural regeneration of annual legumes has been less reliable. Farmers have had to resow legume pastures more often than in the past, or increase their use of nitrogen fertilizer to maintain crop yields.

More intensive cropping due to the new economic pressures in the 1980s has made farmers concerned about the effect of these new practices on soil structure and soil fertility. Moreover, there has, since 1986, been a lift in prices for livestock products and this has resulted in more emphasis and greater interest in sowing legume pastures.

Key Features of the Ley Farming System

Specific Benefits

a. Greater farming stability as farmers are able to diversify and grow field crops and graze livestock providing a more balanced farm income.
b. Increased herbage growth and better quality dry feed in summer resulting in increased livestock production from high protein legume pasture residues.
c. Improved soil fertility and soil structure.
d. Increased cereal crop production and greater cropping flexibility.
e. More effective control of soil erosion, particularly when combined with contour banking in areas susceptible to water erosion, and maintenance of soil cover to prevent wind erosion in drier enviroments with lighter soils.

Main Technical Factors in the Development of the System:

1. Selection and development of annual legumes, mainly *Medicago* spp., suited to a wide range of rainfall and soil conditions.
2. Harvesting methods and a capacity for adequate annual legume seed production.
3. Selection of suitable Rhizobia for effective nodulation.
4. Crop rotations to achieve maximum regeneration of the annual legumes.

5. Shallower tillage systems and tillage and seeding machinery to achieve optimum seed bed preparation and seeding efficiency.
6. Grazing systems that obtain optimum utilization of feed available and seed set of annual legumes.
7. Seeding techniques for effective annual legumes.

Adoption

1. A lift in the profitability of livestock production enabled the continuous cropping cycle to be broken without affecting the farmers' economic viability. This occurred when medic seed was widely available, although supplies were limited.
2. The close liason between research and extension people ensured that the information generated from the research and investigation programs was transferred to the farmers.
3. The widespread use of field demonstrations covering many aspects of ley farming on farmers' properties and at regional research centers, enabled farmers to understand the increased levels of production that might be achieved. Not all the benefits of ley farming are immediately visible, but rather accumulate slowly, a factor that can make adoption more difficult (Rogers and Schumaker 1971). This requires the extension agencies (and the governments that fund them) to be persistent over a period of time.

Relevance of the Ley Farming System to Other Parts of the World

There is increasing evidence that there are large areas of cereal lands in the Mediterranean region where integrated systems of farming can be adopted with substantial benefits (Oram 1977).

The physical and economic conditions in West Asia and North Africa are a mirror image of those that existed in southern Australia in the late 1940s when the ley farming system was introduced, for example:

- livestock products are in demand and their prices high in relation to cereals.
- cereal yields, especially in dry areas, are either declining or becoming increasingly dependent on the use of nitrogen fertilizer.
- soil erosion is widespread.
- both farmers and governments are becoming concerned about land degradation and desertification.

As was the case in Australia, the introduction of annual legumes into farming systems is a potential answer to these problems in the cereal zones.

Many workers have reviewed the potential for ley farming in North Africa and West Asia. Doolette (1980) postulates that taking into account environmental restraints and existing patterns of ownership there may be 15 million hectares in West Asia and North Africa where the system may be adopted with profit.

Carter (1975, 1978) estimated from his studies in the region that, although some technical adjustment would be necessary, the area could amount to 40 million hectares, and if much of the fallow land in countries of North Africa and West Asia were sown to legume pasture, that land could carry many millions more sheep.

There have been a number of programs attempting to introduce ley farming into the Mediterranean basin, in Tunisia, Algeria, and Libya, and more recently in Jordan, Iraq, Iran and Syria.

Although success was achieved on the demonstration areas and farms and on development projects, there has been little grass roots adoption of ley farming by farmers (Oram 1977). The reasons are complex and many technical and social factors are involved. Cocks (personal communication) summarizes some of the reasons:

- the opportunity to involve farmers has been limited.
- Australian medic cultivars selected and developed for mild winter environments have not been highly successful in parts of West Asia or the plateau of North Africa.
- medics have often failed to nodulate.
- in very few cases have livestock been sufficiently involved except superficially, yet it is in its ability to increase livestock production that ley farming is most attractive.
- there has been insufficient follow-up of projects once the initial phase has been completed.

Technical Restraints

Modification of the ley farming systems as practised in southern Australia has been necessary to accommodate the unique climatic, edaphic and sociological conditions found in the region of West Asia and North Africa.

Ecological Considerations

A general comparison of climatic and edaphic conditions of cereal zones in South Australia and areas of West Asia and North Africa reveals some differences. These parameters may be summarized in Table 2. North Africa and South Australia share similar climatic and ecological conditions, although the altitudes in parts of the cereal areas are somewhat higher than those in the South Australia cereal zone.

Identification of Adapted Cultivars

It is now known that Australian cultivars are adapted to climatic conditions occurring in southern Australia, and *M. truncatula* and *M. littoralis* are restricted to regions with mild winters. *M. rigidula*, *M. rotata*, and *M.*

Table 2. Comparative ecological framework of the Mediterranean region and South Australia in the cereal zone.

	South Australia	Mediterranean Region
Rainfall (mm)	300–500	300–900
Rainfall distribution	70% in growing season April-November	Greater winter incidence in growing season
Altitude (m)	0–510	0–1100
Winter temperature (C) Mean min – coldest month	2–10	0–10
Summer temperature (C) Mean max – hottest month	27–32	29–43
Soils	Mixed alkaline and neutral; sand to loam in texture	Mainly alkaline; loam to clay loam in texture

aculata are widespread in cold, dry areas in their native habitat (Fig. 4). Survival of these species through intense frost is naturally superior to the Australian cultivars and under these conditions they produce more herbage, and are well adapted to ley farming practices. In West Asia, *M. rigidula* and

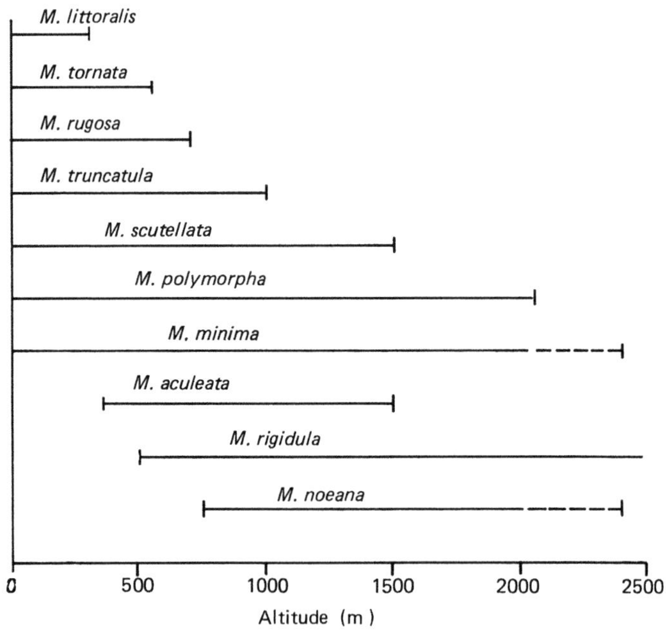

Fig. 4. The general distribution of annual *Medicago* species in relation to altitude in their natural environments (Webber et al. 1986).

M. rotata are best suited to red soils and grey basaltic soils (Cocks, personal communication). A need to use native species is now recognized, and it seems likely that *M. noeana* and *M. polymorpha* are better adapted to alluvial soils, and *M. aculata* to the high plateaux of Algeria. Use of these species overcomes one of the main constraints to the success of ley farming in the region.

Use of Adapted Rhizobia

Further research is needed to find other strains and to study host/Rhizobia relationships. The extent to which inoculation and seed coat pelleting is necessary needs further research but ICARDA has found that seed which has been lime-coated and inoculated with WSM 244 (a strain isolated in Northern Iraq) is normally effective (Cocks 1983).

Conclusions

The ley farming system using annual self-regenerating pasture legumes in rotation with cereals has a number of key interacting technological inputs which must be practised together for the system to work successfully. The system which has developed in southern Australia over almost four decades has accrued significant benefits to agricultural productivity. In theory the system can be adapted to ecologically appropriate areas of the Mediterranean basin but clearly not without modification. The practical adoption of the technology to new areas has been remarkably successful, given the short time involved and the modification to technology that has been necessary. However, developments have often been piecemeal without appropriate attention to the changing of all the key technological inputs, whether they are tillage, legume species, effective *Rhizobium* spp., weed control, timing of operations or grazing management.

To make the system work successfully, it is necessary that adapters of the system have a clear understanding of all components of the technology, and recognise that all components must be appropriately modified and integrated together. They must know that this activity cannot be achieved without substantial technical and extension effort.

Historically, in southern Australia, the ley farming system was developed by farmers and scientists working together and it was implemented primarily by farmers with advice from skilled extension officers. There needs to be greater farmer involvement in future projects to ensure the long term acceptance and adoption at farmer level. The current ICARDA forage program has developed a methodology involving farmers, using their existing mechanical resources. One of the major restraints in the rapid expansion of the program is lack of seed of the adapted cultivars. As was the case in southern Australia, a critical objective of any program must be the availability of adequate supplies of seed and the need to produce seed in the region.

Expansion of the program also needs the technical training of extension specialists and development of extension programs to teach farmers the practical skills of ley farming.

South Australian experience shows that future programs aimed at achieving adoption of the ley farming system in Mediterranean regions need to be based on the following key factors:

1. Early identification of plant cultivars adapted to specific environments;
2. availability of adequate seed and the development of a capacity for seed production of these cultivars in the region;
3. farmer involvement in all phases of the project;
4. greater emphasis on the extension education components of the project.

References

Carter, E.D. 1975. The potential role of integrated cereal-livestock systems from southern Australia in increasing food production in the Near East and North African region. FAO/UNDP Workshop Report. Rome, Italy.

Carter, E.D. 1978. A review of the existing and potential role of legumes in farming systems in the Near East and North African Region. Report to International Center for Agricultural Research in the Dry Areas, (ICARDA), Aleppo, Syria.

Cocks, P.S. 1983. Agronomy Research Report. Agro-Pastoral Development Project, Erbil, Iraq. July 1983. SAGRIC International, Adelaide, Australia.

Crawford, E.J. 1962a. Barrel 173: A new strain of medic for the wheat belt. Journal of Agriculture, South Australia.

Crawford, E.J. 1962b. Cyprus Medic, Leaflet No. 372, Jan 1962, South Australian Department of Agriculture.

Crawford, E.J. 1962c. Harbinger Medic for the lower rainfall wheat belt, Journal of Department of Agriculture, South Australia, February 1962.

Day, H.R., and Michelmore W.A. 1952. Barrel Medic for low rainfall areas. Journal of Department of Agriculture, South Australia November 1952.

Donald, C.M. 1981. Agriculture in the Australian economy (D.B. Williams ed.) Sydney University Press, Australia.

Doolette, J.B. 1980. The Australian ley farming system in North Africa and the Middle East. *in* Proceedings of International Congress in Dryland Farming, Adelaide, South Australia. Department of Agriculture, Adelaide, Australia.

French, R.J. and Schultz, J.E. 1984. Water use efficiency of wheat in a Mediterranean environment. 1. the relation between yield, water use and climate. Australian Journal of Agricultural Research, 35: 743–64.

Hutton, E.M. 1934. Important pasture plants of South Australia. Journal of Department of Agriculture, South Australia 38: 336.

Macindoe, S.L. 1975. History of production of wheat in Australia. Pages 99–121 *in* Wheat and Other Temperate Cereals, (Lazenby, A. and Mathison, E.M. eds) Angus and Robertson, Sydney, Australia.

Oram, P. (1977) Proceedings of an International Symposium on Rainfall Agriculture on Semi Arid Regions University of California, Riverside.

Rogers, E.M. and Schumaker, F.F. 1971. 'Communication of Innovations', Coller-MacMillan, London, U.K.

Trumble, H.C. 1939 Barrel medic, (*Medicago tribuloides*, (Desr.), of Agriculture, South Australia 42: 953–958.

Webber, G.D. 1975. The ley farming system in South Australia. Special Bulletin No. 20/75, Department of Agriculture and Fishers, South Australia.
Webber, G.D., Boyce, K.G., and Simpson, G.H. 1986 Dryland Farming Systems with particular reference to annual pasture legumes in cereal rotations. Technical Bulletin. SAGRIC International, Adelaide, South Australia.
Webber, G.D., Cocks, P.S. and Jefferies, B.C. 1976. Farming System in South Australia. Department of Agriculture and Fisheries, South Australia.

Discussion

Jones:
Farm size and human population density are very different in this region from Australia. How do you see this fitting into the development of a medic- or any legume-based system?

Webber:
I don't believe farm size has much to do with the system. If the farmer has the right seed, and it nodulates and grows properly, he will increase his livestock production.

Cocks:
The critical thing is that the farmer has livestock, otherwise introducing pastures is irrelevant. Farm size doesn't matter at all.

Jones:
Farm size does matter when the security factor is taken into account. The smaller the farmer the greater the risk. It is also relevant to say here that in high rainfall areas (>350 mm) livestock production is only secondary to arable cropping.

Buddenhagen:
1. What would happen if you suddenly couldn't use herbicides?
2. What proportion of south and west Australia is subterranean clover and what proportion is medics?
3. If you applied irrigation to these plants what would it do to their life cycle and terminal growth?

Webber:
1. We would get lower yields. However, the occurrence of the most limiting weeds, wild oats and rye grass, is reduced by proper grazing management, by controlling seeding.
2. In South Australia sub-clovers are 20% and medics 75–80%. In West Australia it is the other way around.
3. In fact, irrigation is only used for high-value crops such as horticulture and seed production. These pasture legumes might last for three or four more weeks.

Springborg:
One of the possible obstacles to the transfer of the Australian technology is the development of machinery. At one end you have the continuation of the manufacture of traditional machines, which have changed very little in recent times, and at the other end the highly capital-intensive importation of equipment. There is no middle range of local creation and adaptation of machinery for new agricultural techniques. We have talked about the need for appropriate machinery, but there has been no definition of what that appropriate machinery might be; so we are modifying imported machines, which may well not be the best thing to do.

Webber:
Maybe we should be encouraging farmers to produce their own seed – they will sow a lot more if they don't have to pay for it – using the simple methods that Australian farmers began with 20 years ago, such as brooms and sheepskin rollers. It must be said that it took 20 years to develop the sophisticated machinery and seed production techniques that exist today.

Haddad:
How much of the impact demonstrated in your data on cereal production is actually a result of ley farming, and how much a result of improvements in cereal production? Do cereal breeders and agronomists agree that all this is a result of the ley farming system?

Webber:
No! If you take ley farming as a total system, a whole array of factors are involved: some effects are due to alleviation of soil erosion, or more effective weed control and fertilizer usage. One of the things that isn't a very significant factor is new cereal varieties, but that never accounts for much in the dry areas.

The Use of On-Farm Research as a Method of Extending Legume Production in Mediterranean Farming Systems

W. ERSKINE, T.L. NORDBLOM, P.S. COCKS, M. PALA and E.F. THOMSON

ICARDA, P.O. Box 5466, Aleppo, Syria.

Abstract. The targeting and designing of on-farm research is described here, with emphasis on the continuous two-way exchange of information about needs and difficulties, new knowledge and techniques, between farmer, extension worker and research scientist. Three case studies are presented: winter-sowing of chickpea in high rainfall areas; mechanical harvesting of lentil in moderate rainfall areas; and the replacing of fallow by forage legumes in low rainfall areas. Results confirmed that adoption of these new technologies would be limited by the high cost of inputs, but that otherwise farmers were sufficiently interested in, and capable of, trying them.

Introduction

The proceedings of this workshop will leave little doubt as to the value of legumes in Mediterranean farming systems: forage and food legumes in most countries, and pasture legumes to replace fallows and to improve marginal lands throughout the region. The achievement of improvements in their production, however, requires the development of a partnership between researchers, extension workers, and most importantly, farmers. This is especially important where new systems are involved, but is true also for the simple introduction of new varieties: this apparently simple intervention has implications far beyond those envisioned by plant breeders.

How should we approach the problem of implementing the recent exciting achievements of research? Clearly we need to know how to test the value to farmers of new technologies, and how to assist farmers in their implementation. We need to know what farmers perceive to be the purpose of new technologies; in what conditions they will be useful; how much they will cost and what the profit will be; and what the internal and external constraints to adoption are. The answers to most of these questions may be found through on-farm experimentation, where two-way flows of information are set up between researchers and extension workers and farmers.

On-farm research is part of the overall 'farming systems approach' (Gilbert et al. 1980; Sands 1986; Collinson 1987) which involves farmers at most levels.

A.E. Osman et al. (eds.), The Role of Legumes in the Farming Systems of the Mediterranean Areas, 273–282.

On-farm Research Methods

Targeting On-farm Research

On-farm research may be designed to meet different objectives, depending on what is needed; diagnosis, testing, or demonstration. This paper focuses on trials aimed towards the introduction or improvement of legume production in targeted farming systems. Targeting involves deciding (a) where to work, (b) which farmers should be involved, (c) how representative they are of the community as a whole, (d) what the current farming practices are, and (e) the nature of the constraints. Brief, preferably informal, surveys can resolve these questions and decide whether different experiments are needed in different agro-ecological conditions.

Designs of On-farm Experiments

Designs tend to vary from the relatively simple designs associated with livestock to more complex designs for agronomic and varietal testing. Two types of on-farm experiments are envisaged: type 2 research in which the researcher predominates in management (scientist-managed on-farm trials), and type 3 research where the farmer predominates (farmer-managed on-farm trials). (Type 1 research is research conducted on research stations). If we take varietal testing as an example, type 2 is very similar to on-station research in that combinations of treatments can be used in small plots, although farms are usually (but not always) used as replicates. In type 3, plots are larger, treatments are usually only two or three, and interactions are not usually measured. Because farmers manage them, they are ideal for demonstration to other farmers, and farmers themselves can assess their practicality.

The design of on-farm experiments with pasture and forage legumes (which should involve livestock) must always be as simple as possible. As will be seen later, these usually continue for several years as account must be taken of their effects on other farm practices, and because responses may vary from year to year due to weather conditions.

Measurement includes the data necessary for economic and statistical analysis and should include monitoring of the farmers themselves, other farmers in the community, and their reactions to the new technology. Especially with type 3, it should be possible from feed-back from farmers to detect fatal flaws in new technologies, and for researchers to see ways to correct these in further trials with altered design. Reference will be made to this last point later in the paper.

External Constraints

The unavailability (or high price) of inputs is the most common external constraint. Some constraints can be overcome by government (seed) or private (farm supplies) intervention, while others (roads, credit, marketing facilities) will require the development of new infrastructure. Some problems (including, on occasions, the above) can only be solved by the development of new government policies. The process of on-farm research, therefore, supplies feed-back on the needs for solutions to external constraints, and for the development of more appropriate technologies.

In the remainder of this paper we present three case studies of on-farm research to facilitate the introduction of legumes. We have chosen examples from three agro-ecological zones, each of which requires the introduction of several concepts simultaneously: (i) the winter-sowing of chickpeas in high (>350 mm) rainfall areas; (ii) the mechanical harvesting of lentils in moderate (300–350 mm) rainfall areas; and (iii) the sowing of forage legumes in low (250–300 mm) rainfall areas.

Winter Sown Chickpea

Yields of chickpea on farmers' fields are low compared with yields obtained using genotypes and improved agronomic practices on research stations (Saxena 1984; Keatinge and Cooper 1984; FLIP Annual Report 1986; FSP Annual Report 1986). Local varieties are susceptible to frost and *Ascochyta blight*, a disease which is prevalent in winter. In the Mediterranean chickpea is therefore sown in spring and is dependent on moisture stored in the soil.

The possibilities that sowing chickpea in winter would make better use of available moisture and make harvesting earlier were first pointed out by Hawtin (1975). Winter planting would result in the development of plants capable of supporting a bigger reproductive structure under conditions of lower thermal stress and better moisture regimes, leading to increased productivity and greater water use efficiency (Saxena 1984; Keatinge and Cooper 1984). At Tel Hadya, progressively advancing the sowing date from March 11 to February 13, December 19, and November 20, gave corresponding increases in yield of 69, 155 and 250% (Saxena 1984). However, for winter sowing, cultivars tolerant to frost and *Ascochyta blight*, better weed control methods, and new fertilizer strategies needed to be developed. By 1984 the package of cultivars (Ghab 1 and Ghab 2) and technologies were ready for testing on farmers' fields.

Preliminary (type 2) trials preceded type 3 testing of the two cultivars with a simple package of sowing date and weed control. In the type 2 on-farm trial results, partitioning of the effects of different treatments was possible (Table 1) (FRMP Annual Report 1987). Winter compared with spring sowing had the greatest effect, while fertilization with phosphorus also gave consistent yield increases. Chemical weed control was superior to no weed control, but inferior (in terms of yield) to hand weeding. However, the unavailability of labour is likely to limit the use of hand weeding. Use of a seed drill (compared with broadcasting) resulted in yield increases, more so with early than with late winter sowings. Interviews of farmers participating

Table 1. Main effect and first order interaction of chickpea yield responses in Syria (kg/ha).

Treatment[1]		1985/86[2]	1986/87[2]	
Time of Sowing	(Early) Dec 85	1409**	Dec 86	1797**
	(Late) Mar 86	1168	Jan 87	1335
Inoculation	(+)	1332*	1538	
	(0)	1265	1594	
Weed control	(chem)	1387**	1498**	
	(other)[1]	1196	1634	
P_2O_5 (kg/ha)	50	1348**	1660**	
	0	1229	1471	
Sowing method	(SPP)[3]	NA	1641**	
	(broadcast)		1490	
LSD (0.05)		67	54	

First order interaction

	Weed control +	Weed control −	Sowing method Bdcst	Sowing method SPP
	**		**	
Time of sowing (Early)	1572	1246	1661	1933
(Late)	1227	1149	1320	1350
LSD (0.05)	94		76	

[1] No control (1985/86), hand weeding (86/87).
[2] * $P<0.05$.
** $P<0.01$.
[3] Single pass planter.
NA = not available.

in the type 2 trials revealed considerable interest in the technologies, but as farmers pointed out, the unavailability of herbicides and seed of the new cultivars are constraints at present.

In type 3 trials, in cooperation with the Syrian Ministry of Agriculture, two winter cultivars were sown by nine farmers, on plots of one to ten hectares, to be compared with local spring varieties and practices in the adjacent fields. The results clearly demonstrated the superiority of winter sowing of the new cultivars which yielded 59% (Ghab 1) and 64% (Ghab 2) more than spring sowing of the local cultivars in the 9 locations (Table 2). From cost data supplied by the farmers winter-sown crops gave twice the profit of spring-sown crops.

Mechanical Harvesting of Lentil

Researchers in many Mediterranean countries have indicated that profits from growing lentil have decreased because of the rising cost of hand labour (Hawtin and Chancellor 1979). A detailed survey of 120 lentil growers in Syria revealed the same problem (Nygaard and Erskine 1980, unpublished).

Table 2. Chickpea production budget in the 1986/87 season.*

	Spring chickpea Local	Winter chickpea Ghab 1	Ghab 2
Yields (kg/ha)			
Seed	991	1574	1629
Straw	650	1542	2127
Gross crop value (SYP/ha)**	11 226	18 085	18 982
Costs (SYP/ha)			
Tillage	410	310	310
Seed	1800	1800	1800
Fertilizer	160	160	160
Weed control	NA	1000	1000
Harvesting	1669	3413	3687
Total	4039	6683	6957
Profit	7187	11 402	12 025
Gain of winter over spring crop		59%	67%

* Based on farmer-managed on-farm trials at 9 locations with field-sized plots (>1 ha) of each type (i.e. spring and winter) at each location.
** Conservative market prices in 1987: SYP 11/kg for seeds; SYP 0.5/kg for straw.

Initial observations of machine performance in lentil harvest indicated that there was no universally suitable machine, but that various systems of mechanization were required for different target groups. Accordingly, all promising machines were tested on- and off-station in a range of agronomic conditions (with and without rolling, ridges, or seed drill) and cultivars (lodging and non-lodging) to establish the minimum change from farmers' management required for each machine (ICARDA 1985). The tractor-mounted double-knife cutter bar satisfactorily mowed an unlodged crop, whether sown with a seed drill, or after broadcasting seeds on to ridges flattened with a heavy bar, but gave unacceptably high losses on a traditional ridged crop. The modified combine harvester required good land preparation, drilling, rolling (where there were stones) and a non-lodged crop. The lentil puller worked well on both traditional and modified agronomy (Friedrich 1988). These three machines relate to different target farmers depending on the area of lentil, the local importance of straw, the capability to prepare a flat seed bed, and the availability of labour, seed drills and so on.

Focusing on the tractor-mounted double-knife mower, on-farm trials (type 3) were initiated in 1985 in cooperation with the Syrian Ministry of Agriculture. Long, narrow plots were used with farms as replicates. The loss of grain from mowing a drilled crop compared to the traditional practice of hand broadcasting was 6%, with a straw loss of 19% (Table 3). Grain losses were outweighed by lower harvest costs (FLIP 1986). Farmers from the village were interviewed at harvest time and they responded enthusiastically to the new ideas, asking where they could buy the machine. They were revisited later for further comments and for updating production costs.

Table 3. Results of a comparison between traditional lentil management practices (broadcast followed by hand harvest) and drilled lentils harvested by double-knife cutter bar averaged over 4 locations on farmers' fields in north Syria in the 1985/86 season.

Sowing method	Harvest method	Seed yield (kg/ha)	Straw yield (kg/ha)	Net benefit over variable[1]	costs[2] (SYP/ha)
Broadcast	Hand	998	2312	2785	1785
Drilled	Cutter bar	937	1866	3059	3119
S.E ±		49.0	100.6		

[1] High mechanization cost.
[2] Low mechanization cost.

A lentil cultivar, 'Idlib 1', with 16% more seed yield and better standing ability than the local cultivar, was released in Syria in 1987: a temporary external constraint to lentil mechanization will be lack of seed of this new cultivar. Of greater concern is unavailability of the machinery. This year the Syrian General Organization of Agricultural Mechanization and private farmers have harvested 10 000 ha by mower and by combine harvester in the Kameshly area, following good seed bed preparation and the importation of mower swathers.

Data collected in surveys indicate that mechanical harvesting should not be evaluated simplistically. Consideration should be given to constraints on the labor available for harvest: mechanization allows much greater areas to be harvested in a limited time frame, a critical point for lentil crops.

Replacing Fallows With Forage Legumes

The objectives of these type 3 experiments were: (i) to measure the effect, on whole-farm productivity, of replacing fallows with common vetch (*Vicia sativa*) and common chickling (*Lathyrus sativus*); (ii) to measure, on farmers' fields, the growth rate of lambs and milk production of ewes grazing vetch and chickling in spring; and (iii) to compare the returns from grazing with those from seed and straw. Unlike the previous cases these experiments attempt to account for all farm enterprises (including the cereal phase of the rotation), and are also complicated by the need for grazing animals.

Before the experiments started, the farmers were interviewed to get background information about the way they fed and managed their flocks, farm and flock size, and crop rotations. At the conclusion, open-ended discussions were held to get their views about forages and the constraints they saw to sowing forage crops.

Incentives were used as little as possible, to ensure that the farmers collaborated simply because of their interest in the new technology, and its potential benefits to them. From ICARDA farmers received seed, phosphate fertilizers, 100 kg of wheat at the end of each season, and medicines for sheep. In the successive seasons farmers increasingly paid the costs of sowing and harvesting.

The project was at Breda (270 mm rainfall), near Tel Hadya, in north Syria (Maertz 1987). In this area barley is grown either continuously on the more fertile soils, or alternating with fallow on the poorer stony soils. The experiments began using relatively small plots and finished by occupying most of the farms of cooperating farmers. An average of seven farms were studied each year, most farms continuing in the project for several years.

Because forages can be used for grain and straw (in the same way as lentil and chickpea), for hay, or for grazing, utilization of the crops was an issue in itself. It soon became apparent that farmers were not interested in hay, so the options were, on the one hand harvesting for straw and grain, and on the other grazing to produce meat, milk, or combinations of meat and milk.

Forage Grain and Straw Yields

Harvesting for seed and straw appeared to be a most attractive option, in spite of the problem of finding labour at harvest time. For this reason farmers preferred to limit the area of forage to less than two hectares, which could be harvested by members of the family. Chickling always yielded more grain than vetch, whether or not phosphate was applied (Table 4). This immediately puts chickling at an economic advantage over vetch and indicates, of course, that chickling may be better adapted to the prevailing poor soils. Farmers reported that chickling was easier to harvest as well. However, there seemed to be little difference in straw yields between the two species, and straw quality was similar.

Lamb Fattening

Fattening appeared to be more attractive to farmers than the grazing of lactating ewes, first, because farmers want to wean lambs early so that they can sell more milk; second, because they need to finish lambs before summer; and finally, because communal grazing, either near the village or in the steppe, provides feed for ewes at no cost. Even though forage yields were low in April, fattening lambs was profitable (PFLP, 1986), but, to achieve reasonable liveweight gains farmers were feeding 400 to 700 g per day per lamb of a barley-concentrate ration. The forages were grazed for about 30 days at stocking rates of 20 to 30 lambs/ha.

Milk Production

Most ewes were in mid-to late-lactation when the forages were grazed in April. Thus milk yields were relatively low – in the range of 400 to 600 g per day, similar to that on native pasture – and unlikely to respond to good quality pasture (Thomson 1984). However, the observed liveweight gain of ewes while grazing is of value because it allows the ewes to recover from the stress of early lactation and thereby enhances the chances of conception (Thomson and Bahhady 1988).

Barley Grain and Straw Yields

Farmers were initially concerned about using fallows to grow forages because they feared that barley yields would be reduced. In the case of barley following unfertilized forage crops there were small reductions in yield (rarely significant), but where barley followed fertilized forages yields were substantially higher: 27% in the case of seed yield and 31% in the case of straw yield. Yields after chickling were higher than after vetch, whether or not phosphate was applied. Although not always statistically significant, if real, these differences are relevant to the farmer in terms of profits: a difference of 200 to 300 kg in straw yield alone would more than cover the cost of the fertilizer.

Table 4 shows that whole farm profitability increased by 50% (vetches without added phosphorus) to 280% (chickling with added phosphorus) as a result of harvesting the forages for grain and straw. Depending on the level of flock management, grazing increased profitability still further.

When interviewed, the farmers were not concerned about possible negative effects of forages on barley, and appreciated the importance of phosphate on both the forage crop and its residual effect on barley. However, they have difficulty finding cash to buy the fertilizer, and the price and availability of seeds were seen as constraints.

Those farmers who had harvested seed in June 1987 were willing to continue growing forages. However if, for any reason, they could not harvest their own seed they stated that they could not afford to buy more seed. In spite of our results they preferred the mature crop option to grazing.

Table 4. Net benefits (SYP/ha) from barley-fallow and barley-forage rotations (BF = barley-fallow; BV = barley-vetch; BC = barley-chickling; 0 = zero phosphorus; P = with phosphorus).

	Rotation				
	B_0F_0	B_0V_0	B_0V_P	B_0C_0	B_0C_P
Gross revenue					
Total per year	1072	1972	3038	2765	3733
Direct costs					
Total per year	429	1005	1279	1005	1279
Net revenue	643	967	1760	1760	2455

Prices and costs (SYP/ha) in November 1986: Forage grain 3.75; barley grain 1.65; all straw 0.80; seed costs = grain prices; fertilizer TSP 1.1. Hand-harvesting costs based on SYP 20 per labourer day and 15 labour days/ha.

Conclusions

The results of all three case studies confirm that the high cost of inputs is an external constraint affecting adoption of the technology. Availability of seed is a second constraint, one which requires direct action from governments, either by encouraging the private sector to produce seeds, or by becoming directly involved themselves. Availability and price of fertilizers, machinery, and labour were also common constraints.

The on-farm research, in addition to highlighting these problems, has revealed that there are few problems associated with farmer expertise, resistance to change, or land tenure. In general, where a technology is seen as advantageous farmers have expressed great interest, and adopted the technology as far as possible. The research has therefore successfully introduced concepts to practising farmers. Extension remains a significant constraint: in most of the cases described the adoption has not occurred beyond the participants of the experimental program.

References

Collinson, M.P. 1987. Farming Systems Research: Procedures for Technology Development. Experimental Agriculture 23: 365–386.

FLIP (Food Legume Improvement Program) Annual Report 1986. ICARDA-109 En, Aleppo, Syria.

Friedrich, Theodore 1988. Untersuchungen zur Mechanisierung der Linsenernte nach dem Rupfprinzip im Vergleich zu anderen Linsenernteverfahren in Syrien. Dissertation 141. Forschungsbericht Agrartechnik des Arbeitskreises Forschung und Lehre der Max-Eyth Gesellschaft (MEG), Gottingen, West Germany.

FRMP (Farm Resource Management Program) Annual Report 1987. ICARDA-131 En, Aleppo, Syria.

FSP (Farming Systems Program) Annual Report for 1986. ICARDA-108 En, Aleppo, Syria.

Gilbert, E.H., Norman, D.W. and Winch, F.E. 1980. Farming systems research: A Critical Appraisal. MSU Rural Development Paper No.6, Michigan State University, East Lansing, USA.

Hawtin, G.C. 1975. The status of chickpea in the Middle East. Pages 109–116 *in* Proceedings of International Workshop on Grain Legumes. ICRISAT, Hyderabad, India.

Hawtin, G.C. and Chancellor 1979. Food Legume Improvement and Development. Proceedings of a Workshop at Aleppo University, May 1978, IDRC, Ottawa, Canada.

ICARDA Annual Report 1985. ICARDA, Aleppo, Syria.

Keatinge, J.D.H. and Cooper, P.J.M. 1984. Physiological and moisture-use studies on growth and development of winter-sown chickpeas. Pages 141–157 *in* Proceedings of International Workshop on *Ascochyta* blight and winter sowing of chickpeas (Saxena, M.C. and Singh, K.B. eds), ICARDA, May 1981, Aleppo, Syria.

Maerz, U. 1987. Methods to simulate distributions of crop yields based on farmer interviews. ICARDA-117 En, Aleppo, Syria.

PFLP (Pasture, Forage and Livestock Program) Annual Report 1986. ICARDA-111 En, Aleppo, Syria.

Sands, D.M. 1986. Farming Systems Research: Clarification of terms and concepts. Experimental Agriculture 22: 87–104.

Saxena, M.C. 1984. Agronomic studies on winter chickpeas. Pages 124–140 *in* Proceedings of International Workshop in *Ascochyta* blight and winter sowing of chickpeas (Saxena, M.C. and Singh, K.B. eds), ICARDA, May 1981, Aleppo, Syria.
Thomson, E.F. 1984. First experiences with joint managed forage and grazing trials. Pages 234–250 *in* Proceedings of the Third Farming Systems Symposium (Butler-Flora, C. ed.), Kansas State University, Manhattan, USA.
Thomson, E.F. and Bahhady, F.A. 1988. A note on the effect of liveweight at mating on fertility of Awassi ewes in semi-arid North-west Syria. Animal Production 47: 505–508.

Discussion

Nassib:
Why do farmers prefer to get seed and straw rather than go for grazing in barley-legume rotations?

Cocks:
Leaving the crop to mature guarantees an adequate supply of seed; the seed and straw can be conserved for feeding during winter, in a time of feed shortage; and grazing can be obtained from communally owned marginal lands, which are a free source. However, a significant number of farmers are choosing to graze forages than to fatten lambs. Marginal land grazing is used for milk production.

Mawlawi:
How do you choose the farmer for your trials?

Erskine:
First the target area is decided, depending on the technology to be applied. Then farmers are approached: they have to be willing to cooperate, already growing the crop, and preferably have land near the main road to allow easy access in winter. They are compensated if the yields are less than they would have got using traditional methods. We sometimes use the same farmer twice. It depends on the experiment.

Webber:
How much are the farmers involved in decision-making?

Cocks:
As far as possible, but with a completely new technology they need a lot of advice and encouragement. After a few years as they gain confidence they make more and more of their own decisions. As for incentives, we don't rent farmers' land unless the technology is very new. Only if we suspect that yield may be low do we offer incentives: we must be realistic!

Haddad:
What kind of incentives are the farmers offered? I think this is most important.

Erskine:
They are guaranteed the return of the local production practices plot for the whole area used. They may be paid in seed or cash.

Kamel:
It depends on the country. In Tunisia we don't rent the land but put in the inputs and if the farmer gets a lower yield than he would have got using his own methods we pay him the difference. In livestock experiments we provide concentrates in some cases.

Human Constraints in Extending the Use of Forage Legumes in Mediterranean Areas

R. SPRINGBORG

Macquarie University, North Ryde, NSW 2113, Australia

Abstract. The integration of forage legumes into farming systems in Mediterranean areas has been impeded by various elements of human behavior and organization. At the level of the international system, concern among development theorists and practitioners has increasingly been focused towards economic policy leverage and away from technology evaluation and transfer. Additionally, ley farming, which is most highly developed in Australia, is subject to the structural problems of technology transfer from one peripheral region of the global political economy to another. At the level of national agricultural policies, commodity price distortions and inappropriate manpower training constrain the expansion of forage legume cultivation. At the level of local farming systems, the principal constraints are patterns of land tenure, with privately owned farms generally too small to permit replication of classic ley farming and the separation of cropping from livestock raising. Farming systems in the region are increasingly variegated as a result of the modernization and bimodalization of agriculture, hence strategies for the dissemination of legume cultivation will have to be tailored to sharply different types of farming systems.

Introduction

The integration of forage legumes into cropping systems in Mediterranean areas has been hindered by a variety of technical factors, including problems of selecting appropriate legume ecotypes and rhizobia, of countering destructive pest attacks, of determining proper tillage techniques in poorly structured soils, and so on. Additionally, the spread of legume cultivation has been impeded by various elements of human behavior. Because more progress has been made in the past decade in overcoming the technical rather than the human constraints, it will probably be the latter rather than the former that pose the more enduring obstacles to extending the use of forage legumes. The purpose of this paper is to identify some of the main impediments resulting from patterns of human organization and to suggest possible means by which some of them may be overcome. The constraints to be dealt with are to be found at the level of international system within

A.E. Osman et al. (eds.), The Role of Legumes in the Farming Systems of the Mediterranean Areas, 283–294.

which agricultural technology transfer occurs; as a result of national agricultural policies; and at the level of local farming systems.

Constraints Imposed By the International System of Technology Transfer

Development theory has passed through several distinct phases since that term came into vogue after World War II. At present the prevailing orthodoxy of development theory is that an appropriate national economic policy is the prerequisite for development. Earlier concerns with building institutional and human resource capabilities and with achieving equity have been downgraded. Social scientists and development practitioners associated with bilateral and multilateral aid agencies now share the opinion that distortions introduced as a result of controls over key productive sectors, and as a result of governmental interference in the economy more generally, are the principal obstacles to more rapid economic growth. The recommended solution, therefore, is for export-led growth to be facilitated by (a) renewed emphasis on prices and markets, (b) a corresponding downgrading of non-market-based allocation systems, (c) currency devaluations and removal of other barriers to trade, and (d) shifting investment from the public to the private sector.

The relevance of this new orthodoxy of development to extending cultivation of forage legumes is that important bilateral and multilateral development assistance agencies have downgraded their interest in specific technologies of production. Since the primary orientation of these agencies is that appropriate economic policies rather than the absence of modern technologies have been responsible for inhibiting agricultural growth, the focus for their activities increasingly is centered on economic policy making. The operating assumption is that new technologies and systems of production in agriculture or elsewhere will not have the desired effect in the absence of policies predicated on private sector-led export growth. Increased concern with policy leverage has resulted in a decrease in interest in assessing and disseminating specific technologies, of which integration of cropping and livestock raising based on annual self-seeding legumes, a practice known as 'ley farming', may be one. While it could be the case that this or any other agricultural system may not stimulate production increases without supportive national economic policies, it is also true that in the absence of an appropriate dry farming system, rainfed agriculture is unlikely to expand rapidly, regardless of the economic regime under which it operates. While such a system may be adopted even in the absence of active support from development assistance agencies, if there were such support, the process presumably would be hastened.

A second obstacle confronting the transfer of ley farming technologies to the Middle East and North Africa is that the primary source of those technologies is Australia, which is more peripheral than central in the global political economy. The transfer of technology from one peripheral region to another confronts difficulties resulting from the historically dominant role of the global center (Europe and North America), and from comparative inadequacies in technology transfer infrastructure on the periphery.

Since the onset of colonialism, the Middle East and North Africa have turned to the West for sources of technology. The introduction of deep ploughing from Europe is but one example of numerous inappropriate technologies that were transferred. Appropriateness, in fact, is only one of the elements determining the selection of technologies that are transferred. Because the global center possesses a highly developed infrastructure through which its technologies can be transferred, it has a competitive edge. Provision of development assistance with the requirement that it be spent on goods and services provided by the donor country, education of Third World students in European and North American universities, and the domination of information gathering and dissemination, all tend to reinforce the paramount position of American and European agricultural technologies. Among Arab agronomists the influence of the world center is reflected by the prestige attached to European and North American university degrees, whether in appropriate subjects or not. Such credentials are far more common and carry more prestige than degrees in aspects of dry land agriculture from universities in Australia, Latin America, or the sub-continent. Governmental institutions in the southern and eastern Mediterranean areas for formulating and executing agricultural policy are staffed by individuals more familiar with methods that evolved in the dissimilar and more favorable environment of the North Atlantic basin than with techniques of dry farming elsewhere in the world.

Sources of new technologies, if they are on the global periphery, confront special problems in transferring those technologies, as the case of Australia suggests. As a relatively primitive economy by North American and European standards, Australia is not a major provider of development assistance. Annual governmental aid to dry farming technology transfer in the decade of the 1980s has been less than \$500 000. Moreover, Australia generates comparatively little information about its own technologies. With regard to ley farming, it was not until 1976 that a publication outlining its operations and management was available (Webber et al. 1976). There is still no book on the topic. Other than the annual reports of the overseas projects departments of the South and West Australian governments, there are no systematic reviews of dry farming projects with Australian involvement. Even if the world's information system were not attuned primarily to the dominant sources in North America and Europe, it would still be difficult for Australia to publicize its technology, because it generates insufficient information about it.

Global information exchange, in sum, is more heavily conditioned by the structure of the world's political economy than it is by other considerations, including those of analogous climatological regions in the northern hemispheres. This impediment to the dissemination of appropriate technologies is being reduced as a result of the activities of multilateral agencies, including ICARDA in the case of dryland farming, but it still remains an important one.

Constraints Imposed By National Agricultural Policies

National agricultural policies in the Middle East and North Africa, as elsewhere, are based on and supportive of prevailing farming systems. It is not surprising, therefore, that attempts to modify those systems or introduce completely new ones run into conflict with practices and policies institutionalized at the national level. Price structure for agricultural inputs and commodities is the most critical area, in part because throughout the region prices are heavily influenced by governmental intervention.

With regard to extending the use of forage legumes, several specific aspects of price structures common to countries in the region have been found to present complications. Subsidization of competitive fodder sources is a major difficulty. In Jordan, for example, the price of barley is so heavily subsidized that a very low ceiling is imposed on the maximum price for forage legumes or hay. The *reductio ad absurdum* of subsidies is in Egypt, where bread, costing .01 USD per loaf, is frequently fed to livestock. Free utilization of state-owned lands in marginal areas for opportunistic cereal crops by enterprising, but resource-degrading, farmers is another hidden subsidy.

A subsidy directly competitive with the extension of legume cultivation is that of nitrogenous fertilizer, because organic nitrogen fixation is one of the key benefits of those legumes. As more fertilizer plants come into production in the oil-producing states of the region, the likelihood of continued and even increased price support for nitrogenous fertilizer grows accordingly. The use of inorganic fertilizer does little to improve soil structure, while the benefits of improved soil structure as a consequence of legume cultivation are largely unknown and unappreciated by the region's farmers. On the output side, the heavy subsidies for locally grown wheat make it more difficult for any competitive crop to be adopted. Since legumes compete directly with wheat in those cereal-growing areas where fallow is not used, or is used only intermittently, and indirectly (at least initially), where legumes are grown in place of fallow, an artificially high price for wheat poses a serious threat to legume cultivation.

Another aspect of national agricultural policy that constrains the use of pasture and forage legumes is the way in which agricultural expertise is structured and trained. Animal husbandry is invariably separated from agronomy, which poses problems for the adoption of farming systems based on the integration of livestock with cropping. The poor performance of most departments of agricultural extension in the region is also an obstacle to the dissemination of new technology. For various reasons governmental linkages between research station and university campus, on the one hand, and farmers on the other, are weak. This problem is compounded in the case of legume cultivation and ley farming by virtue of the fact that most of the region's agronomists have been trained in northern hemisphere universities, or in local institutions whose curricula reflect those of North American or European universities.

However, although most national agricultural institutions and policies are at the very least not strongly supportive of the extension of forage legumes in farming systems, they do not pose insurmountable barriers. Various modifications of traditional ley farming systems can be made to accommodate specific policies. If, for example, wheat production is heavily subsidized, then the benefits of nitrogen-fixing legumes for subsequent grain crops may provide a key to their dissemination. If fodder markets are highly developed then it may be economical to make hay. In general, the high price of livestock in the region suggests that forage legumes may be utilized as permanent pasture rather than being rotated with wheat or barley, as is the case with classic ley farming.

Constraints Imposed By Farming Systems

Land Tenure

Since the early 1970s disenchantment with large scale socialist and even communist landholding patterns (e.g., cooperatives, collectives, and state farms) has caused most Arab governments to break up these holdings and distribute land to individual proprietors in medium-sized units (Springborg 1977). These farms vary in size by country and by district. In northwestern Iraq, for example, they can be as large as 125 ha; in the Jabal al Akhdar region in Libya, which receives a higher rainfall, they are some 40 ha; while in dry farming areas in Syria farming units tend to be somewhat smaller. At the bottom end of the scale there has been some consolidation of the smaller, uneconomic holdings which were created by the agrarian reforms of the 1950s and 1960s. As a result of these two trends the most rapidly expanding type of farming unit, and that which accounts for the largest share of dryland cropping, is the medium sized, privately owned farm, which in general tends to be considerably less than 100 ha. These holdings are very much smaller on average than the cooperatives and state farms on which experiments were conducted with forage legumes in the 1970s and early 1980s. If those legumes are to be integrated into a viable farming system on these privately owned holdings, they will probably have to be divorced from the broadacre farming systems in which they are utilized in Australia and elsewhere.

One potential problem in gearing down a farming system based on alternating pasture with crops is the annual provision of grain. In ley farming this is achieved by dividing operating units roughly in half by fencing, so that some 50% of land is given over at any one time to crops and

50% to pasture. On smaller farms this may not be practical because of size limitations or because of the comparatively high cost of subdividing small holdings. On such farms boundary fencing is in any case disproportionately expensive, so internal fencing is probably out of the question.

There are various ways in which size limitations may be overcome, some of which concentrate on modifying existing tenurial arrangements, and others on encouraging suitable on-farm practices. So, for example, farmers can be induced to pool their land and exploit it cooperatively, as is frequently done among members of family units in the region. Alternatively, individual farms may be given over to crop/pasture rotations that annually occupy 100% of the land, so long as the farmer does not own livestock and is content to average income over a two-year period. This drawback may not be critical because a high proportion of landowners are absentee and have other sources of income. In some cases it may prove feasible to divide even relatively small holdings. In Iraq this has been accomplished on several demonstration farms with fencing. In Jordan, experiments have been conducted with shepherds managing flocks on small farms that have borders but no internal fencing.

An associated difficulty with scaling down the broadacre model is suitability of equipment. Most agricultural machinery developed for extensive fodder and forage legume production is large, expensive, and inappropriate for owner-operators in the Middle East and North Africa. While some commercial contractors may profitably operate various pieces of this machinery, owner-operators, by modifying existing equipment and practices, can probably obviate the need for new equipment or for mechanization generally. Indeed, some traditional labor-intensive methods appear to be quite efficient and suitable for legume cultivation. Analyses of planting methods in Syria, for example, suggest that a skillful hand broadcaster can achieve seed distribution and depth rates almost equivalent in consistency to those of modern seeding machines, while semi- and completely mechanized methods currently in use are also comparatively effective (Harvey 1980).

The changes in land tenure patterns resulting from land reforms, and the increasing opportunities for off-farm employment, have led to the growth of contract servicing in the agricultural sector. Such services are now available in many areas of the region for just about all operations from tillage to harvest. Agrarian reform plots have in many countries proved to be too small to support families, a problem compounded by fragmentation through inheritance. This push factor, combined with the pull of expanding economies as a result of the oil boom, has resulted in a vast increase in off-farm employment for farming families. In Egypt, for example, non-agricultural employment now contributes on average more than half an average farming family's income. This development has in turn made necessary and possible the existence of custom service operators, who in regions like northwestern Iraq now account for the overwhelming majority of agricultural services performed.

Absentee and semi-absentee ownership coupled with custom servicing is probably not conducive to innovation generally, and to the extension of legume cultivation more specifically. Contractors and landowners who rely upon them generally pursue minimax strategies, investing a minimum level of inputs to secure a reasonably guaranteed, but relatively low level of outputs (Farming Systems Program Report 1980). A higher risk strategy would require enhanced commitment by both parties and probably closer supervision of contractors by landowners. At present, most operations provided on contract are standardized at minimally acceptable levels. For improved tillage, seeding, weed control, harvesting and other services, special and more costly arrangements have to be made, if they are in fact available. Income expectations of contractors are based on rapidly performed operations on a large number of holdings, whereas the expectations of landowners are of annual supplements to their non-agricultural earnings without much, if any, involvement on their part in farming operations. Moreover, farming systems based on contractors providing services for absentee owners are not easily made compatible with livestock management, which requires constant physical presence in the absence of a well developed extensive system of livestock control.

The introduction of forage legumes into this system, therefore, will require that the minimax strategy commonly pursued by owners and contractors be replaced by strategies of income maximization that involve greater financial commitments and improved performance levels. To achieve these changes, technology transfer efforts may be focused on either owners or contractors, there being some reasons to believe that the latter target audience may be the more receptive. On the whole, contractors are successful, ambitious entrepreneurs, indicating that they may have greater capability and desire to adopt a system that has the potential to provide greater rewards. For this to be accomplished, however, a system whereby they derive adequate returns from their enhanced commitments must be devised.

Livestock Management

If forage legumes are to become widespread in the region, those who invest resources in growing them must gain suitable benefits either directly or indirectly from increased livestock production. This is particularly the case in the drier areas of the region, where livestock production accounts for very high shares of family income (Farming Systems Program Report 1980). At present, legume pastures are generally not recognized as a sown crop, so they are treated as a public resource, available to all on a 'first come, first serve' basis. This is so even in cases where the pasture is surrounded by fences. While there are some integrated crop and livestock farms in the region, they are comparatively rare. A related problem is that livestock grazing, regardless of ownership patterns, tends to be intermittent. Periods of heavy feeding are followed by those of near starvation. Forage legumes, however, require controlled, moderate and consistent grazing rates. Even if those responsible for managing livestock are aware of this requirement it is difficult to impose proper grazing regimes because of the constant threat that the pasture will attract other flocks.

The historical evolution of livestock management in Australia traces a path that may be followed by southern and eastern Mediterranean animal husbandry (Chatterton 1979). It was not until the gold rush of the latter half of the nineteenth century had absorbed so much agricultural labor that Australian landowners were forced to fence their properties. In many areas of the Middle East there is now a growing shortage of agricultural labor. Wages, including those of shepherds, have increased substantially since the mid-1970s. Comparative economic advantage may, therefore, cause shepherds and intensive methods to be replaced by fences and accompanying extensive management techniques.

It would be unwise, however, to assume that this shift will ultimately occur, and probably equally unwise to attempt to force the pace of change prematurely. At present in most parts of the Middle East and North Africa fencing pasture and crop land is atypical, for not only is it comparatively expensive but it conflicts with the customary right of access to crop stubble, pasture, and fallow. Political constraints also restrict the spread of fencing. Infringements of rights to pasture and stubble are simply not politically acceptable. So, while the long-term goal may be to substitute fencing for shepherds and award to landowners all rights of usufruct whether the land is in crop or not, in the meantime other ways of managing sown legume pastures must be found.

Various means may be used to ensure that benefits accrue to those who grow fodder and forage. Convincing landowners to purchase and raise livestock appears to be the simplest way, but it is not without drawbacks. The problems associated with absentee ownership, including a preference for grain crops, low levels of commitment to agriculture generally, and the absence of customary arrangements for livestock management on properties owned by absentees must all be dealt with. The prejudice of current farming systems against integration of cropping and livestock production poses another set of problems. Moreover, that division is not confined to farms, but is also to be found in university departments of agriculture and in relevant government bureaucracies. In all of these settings the institutional and physical separation of livestock management from cropping is virtually total (Chatterton 1979; Pattison 1980).

Selling forage on an agistment (the taking in of livestock for feeding at a specified rate) basis may be another method by which cropping and livestock production can be integrated. Such a system would obviate the need for fencing if animals could be controlled through tagging or some other means of monitoring consumption. Alternatively, fodder crops can be cut for hay, for which there are increasingly well established markets in the Middle East and North Africa. This is, however, a less efficient method than grazing, both because cutting, baling and transport are required, and because fields have to be reasonably level and free of large stones and other obstacles to allow mechanized hay cutting. Moreover, governmental subsidies of competitive sources of animal feed, including barley, frequently make hay making of legumes unprofitable. The problem of subsidies is aggravated in years of low rainfall when governments and international organizations increase support levels through price adjustments and importation. In years of plentiful rainfall public grazing resources may make additional sources of feed superfluous. Given these market uncertainties and comparative disadvantages of fodder production it is unlikely that legume cultivation, if it is to depend entirely on hay making, will spread rapidly or widely in the area (McWilliam 1982).

Finally, traditional norms and patterns of behavior could be utilized to ensure protection of pastures. Undersowing cereal crops with legumes is one such method, for the rows of wheat or barley are recognized as crop and thereby secure the field for the owner. Following the cereal harvest the stubble and legume pasture can be grazed on an agistment basis, as has been done in Jordan (JADFP 1983). Similarly, while shepherds will cut fences to gain access to pasture, they will respect crops sown by transhumant bedouin along their path of migration and to which they intend to return in the spring (Chatterton and Chatterton 1981). Such a strong cultural proscription on grazing sown crops might be utilized in some imaginative way for the protection of forage legumes. Alternatively, the *hema* system, by which specific grazing areas are recognized as resources belonging to particular social units, might be modified to include sown pastures (Draz 1980, 1983).

Conclusion

Although some aspects of the international system of technology transfer and of national agricultural policies are not conducive to extending the use of forage legumes in the Middle East and North Africa, ultimately these constraints may not be decisive. If legumes are indeed as potentially useful for fodder and forage in this region as they are elsewhere and as experimentation in Syria, Tunisia, Libya, Iraq, and elsewhere has suggested, then farmers in the region will probably incorporate them into their farming systems one way or another. Innovation, after all, does not have to be planned and directed from above to succeed. Theories of development, even when embodied in the recommendations with money behind them from aid agencies, do not necessarily determine outcomes at local levels. Moreover, while national agricultural policies take little account of the possibility that forage legumes could play an important role in increasing livestock production, some aspects of those policies coincidentally are quite supportive. The high price for domestically produced livestock is one example. Another may be spreading awareness in governmental circles in the region that increased agricultural production depends on more favorable treatment of the rural sector, including higher prices for agricultural commodities and reduction of controls over producers, and on an enhanced role for markets in determining input and output prices.

Farming systems in the region are undergoing rapid change, a process which has included polarization. On the one hand, modern, capital-intensive agricultural operations are emerging as private entrepreneurs invest in horticulture, livestock production, and even field crops (Springborg 1986). On the other hand, small-scale traditional agriculture is in many areas stagnating, partly because returns to owners of small plots compare unfavorably with their earnings from off-farm employment. While the problems created by this bimodalization of agricultural sectors transcend the issue of extending forage legume cultivation, they also affect the prospect of its occurring. Precisely because the constraints and opportunities affecting farming operations in the modern and traditional sectors are so different, their comparative receptivity to forage legumes is likely to vary. The farming systems into which legumes may be adopted in these two different sectors are likely to be highly dissimilar. Further research should therefore take account of agricultural bimodalization when assessing constraints to extending the use of forage legumes.

Acknowledgement

This paper includes edited portions of materials previously published in 'Impediments to the Transfer of Australian Dry Land Agricultural Technology to the Middle East', Agriculture, Ecosystems and Environment, 17 (1986), pages 229–251. I would like to thank the publishers of that journal for permission to draw on that material. Readers may also be interested in consulting a related piece by the author, Food Production in the Arab States: The Applications of the Australian Dry Land Model. Journal of Arab Affairs: 3, 2 (1984).

References

Chatterton, B. 1979. Report on rainfed cereal and livestock production in West Asia and North Africa. South Australian Ministry of Agriculture and Fisheries, Adelaide, Australia.

Chatterton, B. and Chatterton, L. 1981. Combatting desertification in winter rainfall regions of North Africa and the Middle East. Outlook for Agriculture 39: 397–402.

Draz, O. 1980. Range and fodder crop development: national range management and fodder crop development program, Syrian Arab Republic. FAO AG: DP/SYR/68/001.

Draz, O. 1983. Rangeland conservation and development. World Animal Review (July-September): 2–14.

Farming Systems Program 1980. Research Report No.2. ICARDA, Aleppo, Syria.

Harvey, J.A. 1980. Planting methods for winter crops in NW Syria. Discussion paper, ICARDA, Aleppo, Syria.

Jordan-Australian Dryland Farming Project. Annual Report, 1983. Australian Development Assistance Bureau, Canberra, Australia.

McWilliam, J.R. 1982. Pasture and forage systems in North Africa and West Asia: A proposal for research. Paper prepared for Director General of ICARDA, Aleppo, Syria.
Pattison, R.J. 1980. The need for integration of livestock with cropping enterprises in the *Medicago* base ley farming system in Algeria. The Second International Congress on Dryland Farming, Adelaide, Australia.
Springborg, R. 1977. New patterns of agrarian reform in the Middle East and North Africa. Middle East Journal 31: 127–142.
Springborg, R. 1986. *Infitah*, agrarian reform, and elite consolidation in contemporary Iraq. Middle East Journal 40,1: 33–52.
Webber, G.D., Cocks, P.S. and Jefferies, B.C. 1976. Farming systems in South Australia. Ministry of Agriculture and Fisheries, Adelaide, Australia.

Discussion

Capper:
You gave as an example of deintensification the increased growing of berseem in Egypt. I would suggest that this has been caused by pricing policy in that until recently only livestock products and vegetables were not subject to price controls: all other agricultural products were.

Springborg:
Various USAID studies indicate that preference for berseem cropping is a function of both price and labor demand.

Solh:
Your comment on deintensification of agriculture in Arab countries in North Africa and West Asia does not reflect the true situation if we consider the following examples:

- Egypt. On both sides of the Nile are greenhouses with crops grown all through the year highly intensively, with yield levels in major crop commodities as high as those obtained in California.
- Ghor Valley in Jordan.
- Saudi Arabia. The very successful story of wheat, poultry and dairy products.
- Morocco. Expansion of irrigated area up to about 800 000 ha with plans to reach one million. Production in many major food commodities was increased in most crops, e.g. self-sufficiency ratio in sugar increased from 0 to 65%
- Algeria. Fallow is being replaced by food and forage crops thus intensifying cropping.
- Greenhouses are mushrooming throughout the area and many warm-season vegetables are on sale throughout the year, which was not the case in the past.

Bearing in mind that 90% of the area is rainfed, the generalization of deintensification is very unfair. Besides, the problem of the very high population increase rate might be masking the benefits of intensification.

Jones:
The problem may be the use of the word 'intensification'. I think most agriculturalists use the word to mean 'intensification of cropping'. In the barley system, for example, if a farmer used to grow a barley-fallow rotation and now grows barley every year, that is intensification.

Springborg:
I was using the word to include all aspects of inputs, including labor. Some farmers everywhere follow minimax strategies, investing a minimum of inputs to minimize risk and obtain a lower level of output than could possibly be obtained through more intensive inputs. In the Middle East and North Africa the tendency has been accentuated by two factors: 1) inappropriate macro-economic policies for the agricultural sector, including price controls for commodities, and 2) increased opportunities for off-farm employment. As a result, the traditional agricultural

sector has undergone a process of 'deintensification'. On the other hand, 'ag-investors', frequently from urban areas, have invested a significant amount of capital in feed-lotting, horticulture, etc. This remains a comparatively limited factor in total acreage and output, however.

Cocks:
I am worried that international funding bodies now believe that 'institution building' is not worthwhile, and in particular that extension is now seen as inhibiting the spread of agricultural technology. It seems to me that extension services are an essential link, both ways, between governments and farmers. If there is a lack of avenues of communication between them it will be difficult to get adoption of our research.

Springborg:
While I agree that extension services can play a role in technology transfer, they are only one such agent. Innovative farmers may well account for a larger percentage of techonology adoption than government extension agents. Agribusiness concerns are also very active in disseminating technology and in many if not all regions of the Middle East play a much greater role than the government extension services. Finally, it is possible to use rural credit institutions to disseminate new technologies. Such a program was developed in Pakistan by USAID and is now being applied in Egypt. It involves an assumption of greater risk by farmer clients and clever evaluation of potential for success by loan officials, who frequently possess credentials in agricultural sciences. Extension tends to be equated in people's minds with government control: there is evidence, from Africa, of a correlation between a decrease in government control and an increase in production.

PART V

WORKSHOP DISCUSSIONS AND RECOMMENDATIONS

Summary of and Issues Arising From the Discussions

The workshop discussion sessions proved to be most stimulating and fruitful, yielding many valid comments and interesting points of information not covered in the papers themselves. The editors believe that a summary of the main issues can make a valuable addition to these proceedings. It was not possible to include all the ideas expressed, but even if not recorded they all contributed to the lively and constructive exchanges, much appreciated by the organizers and participants.

Definitions

The workshop found it useful to define certain terms of particular relevance to the design of future strategies for legume cultivation in the farming systems of Mediterranean areas:

1. Mediterranean: except in the first paper, which takes a global overview, 'Mediterranean' applies only to the countries bordering the Mediterranean Sea and those nearby with similar agro-climatological conditions (mild wet winters, hot dry summers).
2. Food and Feed: there is no agronomic division between the so-called 'feed legumes' and 'food legumes'; these terms are simply a reflection of their use, and many legumes are multi-purpose. 'Grain legumes' was preferred.
3. Marginal: means any land that is not suitable or is excessively risky for cultivation, either within the cultivated areas or at the edge of the steppe. The limiting factors are not only low rainfall, but include saline, shallow and/or stony soil, and steep slope.
4. Fallow: should be defined clearly so that 'clean fallow' (land kept clear of plants) is not confused with 'weedy fallow' or 'volunteer pasture'.
5. Zone: as it means different things in different countries, it should always be qualified, i.e. by mention of average rainfall.

Human and Animal Nutrition

The benefits of legume cultivation in the farming system were never in doubt, but the nature of the market for these crops was considered an important issue, because ultimately it is the demand, as well as the agricultural conditions that will determine the type and quantities grown. Nutritionists are well aware of the advantages of a combination of legume and cereal protein in the diet: will people in fact eat more legumes if they become more widely available? Will they change their eating habits to include the new legumes that may be introduced as food crops in the drier areas? The anti-nutritional factors would have to be modified or eliminated,

A.E. Osman et al. (eds.), The Role of Legumes in the Farming Systems of the Mediterranean Areas, 297–304.

through breeding and processing, and people would have to be better informed about the benefits to health of legumes in the diet, so that they don't feel deprived or ashamed about eating them, and choose to buy beans and lentils instead of meat. There was some discussion on whether breeders should concentrate on improving the quality, i.e. the size, colour, shape, taste and texture, to make them more attractive to the consumer and so more profitable to the farmer. (Seed size can make as much as 50% difference to the market price).

Discussions confirmed that in the Mediterranean areas feed resources are the greatest constraint in livestock production, so the greatest demand for legumes is as livestock feed; as hay, silage or grain, grazed green or as stubble, or turned into concentrates. There is no 'best' way to utilize such legumes; each agroclimatic area and each farming system has its own methods. Nutritionally speaking, the best way to use a forage legume is to cut it when mature and use the hay as conserved fodder, but that can't work in a nomadic system, where the owner of the animals does not own land, and where time and equipment are limiting factors. Silage is made in some places, but on the whole suitable technology is neither available nor understood. Milk producers were said not to like silage. Feed for poultry as well as small ruminants was considered. Participants were reminded that there are 20 million chickens in Syria alone. Where is their feed coming from? Lupins could be grown as a substitute for imported soyabean, and some chickpea is already being used, although it lacks oil. Modern livestock systems tend to use concentrates. In Portugal, for example, they are cheaper than straw, but that is an exception in this region. The method of utilization has to be considered by breeders: if straw is valuable then an increase in seed yields is not enough; they should also aim for increased biomass. Fortunately, with legumes the two are closely associated.

It was observed that the question often arises of competition for scarce resources between humans and animals. Participants agreed that to a very large extent small ruminants are in areas or are utilizing crops that could not be used for food anyway. The role of these animals is of tremendous significance in the Mediterranean areas because they use resources otherwise unavailable to man and convert them into forms digestible by humans. These resources are natural pastures, and here the leguminous flora have an important potential role. However, the proposed improvements through phosphate fertilization and controlled grazing will not be possible without considerable research into socioeconomic factors, particularly land tenure.

Water, another basic resource, was not forgotten in the discussions. Better use could be made of rain water by harvesting it to supply flocks in the drier areas.

Crop Sequence and Livestock Interaction

Legumes fit well into the farming systems of the Mediterranean areas because of their flexibility. In such a region, where the seasons are so variable, this flexibility is necessary to optimize the use of resources and to balance risk. However, introducing pastures or a crop simply for feed is irrelevant unless the farmer has livestock, or unless there is the infrastructure to store and transport forage on a large scale, because arable and livestock are not necessarily in the same zone. This prompts the question of transhumance: is it easier to move the animals or the feed? Indeed, it raises the whole issue of integrated farming. The versatility of legumes can only be fully realised in an integrated crop/livestock system, where legumes are grown in rotation with cereals, animals can graze the crop residues and benefit from conserved legume forage in the winter, and the cereals can gain from the legumes' improving effect on the soil. The paper on sheep-rearing systems in the French Mediterranean zones gave a pertinent example and aroused considerable interest in this respect, because of the proposed integration of sheep, annual pasture legumes and vineyards.

The precise effect of legumes on soil was a controversial issue during the workshop. The increase in soil nitrogen after a legume crop may not explain all the benefits to the subsequent crop. Indeed, replacing fallow by legumes may have an adverse effect in some conditions, by depleting soil moisture.

It was generally agreed that the choice of which legume to grow in the rotation will be ultimately determined by the relative market prices of livestock and livestock products compared to grain and straw. Indeed, the decision to grow a legume at all, instead of another cereal crop, or leaving the land fallow, depends on such market conditions.

The ley farming system, a special case of legume-cereal rotation, featured in many of the discussions. It is biologically complex (though simple to operate) and new to the Mediterranean farmer, so will only be adopted successfully if the techniques are fully understood to ensure that the seed store in the soil is preserved, and if local varieties of the medics are used (or any other self-regenerating legume: the system need not be restricted to medics). It was suggested that not too much should be made of the technical difficulties, or potential users may be discouraged. It is only really deep ploughing, 25–30 cm, as practised in some places in North Africa, that harms the system. Depths of 15–20 cm still allow regeneration. Nor should ley farming be described as 'Australian', which might make it sound irrelevant to other parts of the world, when the system can work anywhere, with local modifications. Availability of seed is a serious constraint at the moment, but farmers could harvest their own very simply, until machinery is more widely available.

Socioeconomic Factors and Extension

The two case studies described in Part 4, on ley farming in South Australia and fallow replacement in Turkey, illustrate how fundamental good extension and an understanding of the socioeconomic situation is to the success of a new technology. On the whole, the extension services in WANA are not capable: extension agents may be ignorant about practical farming and the researcher often neglects to find out farmers' conditions and feelings first. Extension should continue throughout the research process, from beginning to end, so that real problems are studied, and practicable solutions developed. Maybe extension and research should be in the same institution, and not necessarily in government departments either. Some transfer of technology may occur through the private sector, through agribusinesses, seed merchants and chemical companies. In the perceptions of some, anything from governments is suspect; it takes away freedom and exercises too much control. On the other hand, private companies may only be interested in profitable enterprises, and some fundamentally important issues, like soil and water conservation and land use will be ignored.

Why is it that extension is successful in some places and not others? One reason may be the status of extension agents. In this region it is too low. In other parts of the world research and extension officers have equal status and equivalent qualifications. Or, the barrier may be at the receiving end, as it were. To be receptive to new ideas farmers must be educated, able to comprehend new techniques and the reasons behind any change, so good universal education, from primary school upwards, is vital. New technology will only be of use if farmers are able to implement it, and perhaps most important of all, want to implement it, because its value has been demonstrated and proven.

It was stressed many times during discussions that advice must be right before it is offered to the farmers, so thorough on-farm testing is essential. Research cannot afford to lose credibility by giving unsound recommendations and allowing mistakes.

Land tenure is another crucial factor, not well enough understood or documented in the Mediterranean areas. It is particularly important for pasture improvement: if the users of the land, the shepherds, are not the owners, who should pay for any inputs needed for improvement, the fertilizer, the fencing, and who should receive the benefits?

One way of ensuring that research is actually relevant and needed is to give farmers some control over it. The workshop was told of an interesting system in Australia, where farmers actually pay for research. 25% of research funding is paid for by levies on crops, and Rural Industry Research Councils have to provide a Five-Year Plan which has to be approved by growers. This is not likely to happen here but the lessons from Australia, of as much consultation with farmers as possible, should be heeded.

In order for research to be more widely applied and adapted by farmers, there should be classification of the trial sites and detailed measurements of environmental conditions taken. Climatic and soil data should be specified, disease and pest incidence reported, and any other circumstances that may affect transfer of results to other locations noted.

No-one is going to change familiar methods unless he is certain of benefitting from the change without risk of loss. That certainty will depend upon (a) how much a farmer can afford to gamble – and the smaller the farmer the lower the stakes; (b) how much government support through credit and insurance schemes and subsidies would cushion against loss, and (c) the degree of understanding by the farmers, on the one hand a function of their own educational level, and on the other of the skill of the extension agents.

Workshop Recommendations

Introduction

The aim of agriculture is to produce food for people, ideally a combination of carbohydrates and proteins. In the Mediterranean areas populations are increasing rapidly and there is simply not enough food produced, especially protein. Food and feed imports may soon be crippling national economies. However, as this workshop has shown, agriculture could be more productive by better utilization of the land resource. This is where legumes in the farming systems can play a key role: if more legumes were grown they could help to close the trade gap, and to fill the protein gap in the human diet; directly, with faba bean, lentil, chickpea and other food legumes; indirectly with forage and pasture legumes to be converted into meat and milk by sheep, goats and cows.

There are two ways in which land can be more efficiently utilized to produce more legumes:

(a) by including grain and forage legumes in the rotations instead of the existing continuous cereal cropping or fallowing. This would have the additional benefits of enriching the soil with nitrogen and organic matter, as well as protecting the soil against erosion.
(b) by increasing the proportion of leguminous flora in the native pasture of the steppe and marginal areas, thus providing richer grazing and extending the area of agriculturally productive land.

The following recommendations made by the participants of this workshop suggest the ways in which research and extension can promote legume cultivation in farming systems in order to increase sustainable food production, the overriding objective of agricultural efforts in the Mediterranean areas.

General

Regional collaboration: There should be more interaction between the countries of the Mediterranean region, including southern Europe, regarding legume production.

Human Nutrition

Legumes in the diet: Farmers should be given appropriate encouragement to grow more food legumes so that the cereal-based diet of many people in this region may be improved. Breeders and food technologists should aim to reduce the anti-nutritional factors in legumes to enhance their food value and acceptability.

Drier areas: Research into legume crops suitable for the drier areas (>300 mm rainfall) is needed to increase availability of plant protein for human consumption.

Legumes and Livestock Feed

Grain legume straw: Further research is required on the chemical composition and nutritional value of legume straws, bearing in mind that they are less important than cereal straws in Mediterranean areas.

Mixed feed: There is a need to develop indigenous grain legume crops for use in poultry and ruminant feeds. The aim should be to produce standardised products, both marketable to animal feed manufacturers and competitive with imported protein sources such as soya bean.

Phosphate: Phosphorus is an important element for both plants and animals. Further research is required on the amount, frequency and economics of phosphate application on marginal lands as a means of improving herbage production.

Range degradation: The implications of increases in the production of leguminous animal feed in the arable sector should be studied in relation to the intensity of steppe and range grazing and its effect on land degradation.

Livestock management: In order to strengthen the integration of forage legume crop production and utilization by animals, improvements in livestock management should form a part of farming systems research.

The Role of Legumes in Crop Sequences

Management packages: In order to provide farmers with effective and timely recommendations, national program research should be of a problem-solving, practical nature. It should seek to provide integrated and inter-related management packages for all crops in a rotation. Special attention should be given to:

(a) techniques of seed-bed preparation for good stand establishment;
(b) weed control measures that integrate any use of chemicals;
(c) techniques of harvest mechanization and post-harvest handling.

Quantification: The value of pasture, forage and food legumes to farming systems needs further quantification within the context of the crop rotations within which they are grown. Topics urgently requiring research include:

(a) nitrogen dynamics and fixation;
(b) disease, pest and weed control;
(c) amount and timing of phosphate fertilization;
(d) economic analysis, including the development and use of whole-farm models.

Rotation studies: Crop rotation experiments should not be limited to two-course rotations but should include three- and possibly four-course rotations. Long-term effects should be included in these studies.

Monitoring: Growing legumes in Mediterranean areas is now relatively uncommon, but was once widespread. It would be useful to monitor what is happening to cereal yields, diseases, pests, weeds and economic returns to the farmer in systems where cereal-legume rotations have been replaced by cereal monoculture.

Herbicides: Lack of chemicals to control weeds in legumes is a problem in the region. ICARDA is urged to strengthen its research to find new chemicals and to conduct collaborative studies on their use.

Genetic resources: There is an urgent need to evaluate the genetic resources of the very wide range of annual pasture legumes of the Mediterranean region and their associated rhizobia, in terms of their adaptation to climatic, soil, and management constraints.

The Role of Socio-economics and Technology Transfer

Socio-economic factors: At both the farm and national level it is essential that socio-economic factors are taken into account at all stages in the development of improved systems for legume production.

On-farm research: This is a means of interaction between researchers, extension workers and farmers and is therefore fundamental in the process of adaptation and transfer of legume technologies to the farmer. More training is required in the techniques of on-farm trials.

Seed production: In developing new technology which requires legume seed for implementation, support must be given for the development of seed production, training in seed production technology and provision of high-quality seed to farmers.

Ley-farming: The impact of ley-farming systems in the countries of North Africa and West Asia should be analyzed and evaluated in view of the current difficulties in adapting and transferring the technology to the farmers.

Information: Because of the widespread interest in ley-farming it is recommended that an information manual be prepared by ICARDA for distribution to interested national scientists.

List of Participants

Those marked 1 are senior authors and those marked 2 are co-authors.

ABOU AKKADA, A.R.[1]	Faculty of Agriculture University of Alexandria Alexandria Egypt.
AL-ANNEY, A.H.K.[1]	Field Crops Department State Board for Agriculture and Water Resources Ministry of Agriculture and Irrigation Baghdad-Abu-Ghraib Iraq
BAHL, P.N.[1]	Indian Agricultural Research Institute (IARI) New Delhi 110012 India.
BEALE, P.	International Center for Agricultural Research in the Dry Areas (ICARDA) Rabat Instituts B.P. 6299 Rabat Morocco
BENBELKACEM, A.[1]	Institut Technique des Grandes Cultures (ITGC) Station Experimentale d'El Khroub Constantine 25100 Algeria
BEN SALAH, H.	Institut National de la Recherche Agronomique de Tunisie (INRAT) 2080 Ariana Tunisia
BOUNEJMATE, M.[1]	Institut National de la Recherche Agronomique (INRA) B.P. 415 Rabat Morocco

BUDDENHAGEN, I.W.[1]
Department of Agronomy and Range Science
University of California
Davis, CA 95616
USA

CAPPER, B.S.[1]
Animal Feeds Section
Overseas Development Natural Resources
Institute
56/62 Gray's Inn Road
London WCIX
UK

Present address:
International Livestock Center for Africa (ILCA)
P.O. Box 5689
Addis Ababa
Ethiopia

COCKS, P.S.[2]
International Center for Agricultural Research in the Dry Areas (ICARDA)
P.O. Box 5466
Aleppo
Syria

DAHMANE, A.B.K.[2]
Institut National Agronomique de Tunisie (INAT)
43 Avenue Charles Nicolle
Tunis
Tunisia

DJEMALI, M.
Institut National Agronomique de Tunisie (INAT)
43 Avenue Charles Nicolle
Tunis
Tunisia

DORDIO, A.M.[1]
Estacao Agronomica Nacional (INIA)
Quinta do Marques
2780 Oeiras
Portugal

DURUTAN, N.[1]
Field Crops Improvement Center
P.K. 226
Ulus-Ankara
Turkey

EL-GHANMI, M.	Laboratoire de Controle de Semences 30 Rue Alain Savary Tunis Tunisia
EL-MONEIM, A.A.	International Center for Agricultural Research in the Dry Areas (ICARDA) P.O. Box 5466 Aleppo Syria
ERSKINE, W.[1]	International Center for Agricultural Research in the Dry Areas (ICARDA) P.O. Box 5466 Aleppo Syria
FALCINELLI, M.[2]	University of Perugia Borgo XX Giugno 06100 Perugia Italy
GINTZBURGER, G.[1]	Institut National de la Recherche Agronomique (INRA) – LECSA 9 Place Viala 34060 Montpellier France
GULER, M.[1]	Field Crops Improvement Center P.K. 226 Ulus-Ankara Turkey
HADDAD, N.I.[1]	Faculty of Agriculture University of Jordan Amman Jordan
HALILA, M.H.[1]	Institut National de la Recherche Agronomique de Tunisie (INRAT) 2080 Ariana Tunisia
IBRAHIM, M.H.	International Center for Agricultural Research in the Dry Areas (ICARDA) P.O. Box 5466 Aleppo Syria

JONES, M.A.	International Center for Agricultural Research in the Dry Areas (ICARDA) P.O. Box 5466 Aleppo Syria
JONES, M.J.[1]	International Center for Agricultural Research in the Dry Areas (ICARDA) P.O. Box 5466 Aleppo Syria
KAMEL, A.H.	International Center for Agricultural Research in the Dry Areas (ICARDA) B.P. 84 2049 Ariana Tunisia
LABIDI, M.	Institut National de la Recherche Agronomique de Tunisie (INRAT) 2080 Ariana Tunisia
MAWLAWI, B.[1]	Directorate of Agricultural and Scientific Research P.O. Box 113 Douma Damascus Syria
MEZNI, M.Y.	Laboratoire des Cultures Fourrageres Institut National de la Recherche Agronomique de Tunisie (INRAT) 2080 Ariana Tunisia
NASSIB, A.M.[1]	Field Crops Research Institute Agricultural Research Centre Giza 12619 Egypt
NERSOYAN, N.	International Center for Agricultural Research in the Dry Areas (ICARDA) P.O. Box 5466 Aleppo Syria

OSMAN, A.E.[1]
International Center for Agricultural
Research in the Dry Areas (ICARDA)
P.O. Box 5466
Aleppo
Syria

PAPASTYLIANOU, I.[1]
Agricultural Research Institute
Nicosia
Cyprus

PODIMATAS, C.I.[1]
Fodder Crops and Pastures Institute
411–10 Larissa
Greece

ROBSON, A.D.[1]
Soil Science and Plant Nutrition
School of Agriculture
University of Western Australia
Nedlands
Western Australia 6009
Australia

SEKLANI, H.[2]
Institut National de la Recherche
Agronomique de Tunisie (INRAT)
2080 Ariana
Tunisia

SNOBAR, B.A.[2]
Faculty of Agriculture
University of Jordan
Amman
Jordan

SOLH, M.
International Center for Agricultural
Research in the Dry Areas (ICARDA)
Rabat Instituts
B.P. 6299
Rabat
Morocco

SPRINGBORG, R.[1]
School of History
Macquarie University
North Ryde
NSW 2113
Australia

TAWIL, M.W.[2]
Directorate of Agricultural and Scientific
Research
P.O. Box 113
Douma
Damascus
Syria

WEBBER, G.D.[1]
Plant Services
South Australian Department of Agriculture
Box 1671 G.P.O.
Adelaide
South Australia 5001
Australia

GPSR Compliance

The European Union's (EU) General Product Safety Regulation (GPSR) is a set of rules that requires consumer products to be safe and our obligations to ensure this.

If you have any concerns about our products, you can contact us on ProductSafety@springernature.com

In case Publisher is established outside the EU, the EU authorized representative is:

Springer Nature Customer Service Center GmbH
Europaplatz 3
69115 Heidelberg, Germany

Batch number: 10607222

Printed by Printforce, the Netherlands